AF443430

Handbook of Pollution Prevention and Cleaner Production

Volume 1

Best Practices in the Petroleum Industry

Handbook of Pollution Prevention and Cleaner Production

Volume 1

Best Practices in the Petroleum Industry

Nicholas P. Cheremisinoff and Paul Rosenfeld

AMSTERDAM • BOSTON • HEIDELBERG • LONDON • NEW YORK • OXFORD
PARIS • SAN DIEGO • SAN FRANCISCO • SINGAPORE • SYDNEY • TOKYO

William Andrew is an imprint of Elsevier

William Andrew is an imprint of Elsevier
Linacre House, Jordan Hill, Oxford OX2 8DP, UK
30 Corporate Drive, Suite 400, Burlington, MA 01803, USA

First edition 2009

British Library Cataloguing in Publication Data
Cheremisinoff, Nicholas P.
  Handbook of pollution prevention and cleaner production:
  best practices in the petroleum industry.
  1. Petroleum industry and trade—Environmental aspects.
  2. Green technology. 3. Best management practices
  (Pollution prevention) 4. Industrial ecology.
  I. Title
  665.5'0286-dc22

Library of Congress Control Number: 2009927237

ISBN: 978-0-81-552035-1

For information on all William Andrew publications
visit our website at elsevierdirect.com

Printed and bound in the United States of America

09 10 11 12 11   10 9 8 7 6 5 4 3 2 1

**Working together to grow
libraries in developing countries**

www.elsevier.com | www.bookaid.org | www.sabre.org

ELSEVIER   BOOK AID
            International   Sabre Foundation

Contents

About the authors

Nicholas P. Cheremisinoff is a chemical engineer with more than 35 years of international business, applied research, and engineering experience across several heavy industry sectors. He has led scores of pollution prevention and cleaner production assessments in major industrial complexes and trained several thousand industry professionals on best practices, waste management, and pollution prevention. He has contributed extensively to the literature of environmental and chemical engineering practices as author, co-author, or editor of numerous books and many hundreds of articles. He received his B.Sc., M.Sc., and Ph.D. degrees in chemical engineering from Clarkson College of Technology.

Paul Rosenfeld is an environmental chemist with over 20 years of experience. His focus is fate and transport of environmental contaminants, risk assessment, and ecological restoration. His project experience ranges over monitoring and modeling of pollution sources as they relate to human and ecological health. Dr Rosenfeld has investigated and designed cleanup programs and risk assessments for contaminated sites containing pesticides, radioactive waste, PCBs, PAHs, dioxins, furans, volatile organics, semi-volatile organics, chlorinated solvents, perchlorate, heavy metals, asbestos, odorants, petroleum, PFOA, unusual polymers, and fuel oxygenates. He received a B.A. in Environmental Studies from UC Santa Barbara, an M.S. in Environmental Science, Policy and Management from UC Berkeley, and a Ph.D. from the University of Washington.

Preface

This is the first in a series of volumes on cleaner production and pollution prevention. The intent of the series is to provide guidance on best management practices, technologies, and approaches to managing environmental aspects.

The term *environmental aspect* (EA) refers to the relevant issue(s) that management needs to address, irrespective of level of abstraction, e.g. waste management, worker protection, compliance, public safety, property damage, global warming, and resource extraction, lack of knowledge about process emissions, toxic material management, and biodiversity. A list that identifies the EAs logically leads to defining the inputs to other actions aimed at their management, which forms the basis for both a strategy and action plans.

Companies that rely on a formal Environmental Management System or EMS apply the EA concept to managing compliance issues. In contrast, other companies conduct their business without an explicit list of EAs. These companies generally tend to lack transparency in the priority settings of their environmental work even though they may have a corporate Environmental Policy statement. All companies really should explicitly identify their EAs because if nothing else it brings clarity and transparency to the organization's management of environmental issues. Among the reasons why transparency is needed are:

- Internal to the company management (both decision-makers and line and function) require it in order to effectively implement corrective actions and action plans, and to make the hard decisions concerning resource and money allocations.
- Internal to the company again, the accounting division needs transparency in order to properly account for environmental expenditures in the bottom line.
- External to the company, shareholders and investors demand this today, more so than at any other time in history.
- Homeland security and emergency responders need access to this information because it can impact on emergency preparedness and responses to environmental catastrophes.
- It can serve as a form of insurance against frivolous claims of wrongdoing or legal suits, or in being named as a potentially responsible party (PRP) for environmental damages.
- It can help protect the future value of assets or property, especially at the time of sale or in mergers and acquisitions.

The term EA is used to identify the important issues in the environment that an organization should take into consideration in their environmental work. These include things that we care about due to individual human aspects, such as noise, odor, occupational exposures to potentially harmful environments and situations, laws and regulations, being a good neighbor and responding to a complaining community. EAs can also be a product's or production process's *environmental impact*, e.g. emissions to a nearby stream, lake or river, stack emissions, and the overuse of energy. Some other examples are emissions of a chemical, waste generation, production leakage, recycling, different materials, hazardous materials, electromagnetic fields, and impact on flora and fauna.

There are different types of EAs according to ISO standards. The EMS, ISO 14001, and other international standards are used by companies to find a common basis for managing the environment affected by a business's operations. Implementation of an EMS like ISO 14001 includes defining environmental policy, planning and implementing of an environmental program, checking measures according to goals, and reviewing by management. ISO 14001 is characterized by demands for continuous measurements and an EMS is business focused.

According to ISO 14001, EAs are "elements of an organization's activities, products or services that can interact with the environment" (according to standard ISO 14001:1996). The environment is defined as "surroundings in which an organization operates, including air, water, land, natural resources, flora, fauna, humans, and their interrelation."

The term *environmental impact* (EI) is "any change of the environment, whether adverse or beneficial, wholly or partially resulting from an organization's activities, products or services." Annex A to ISO 14001 states that the process to identify the significant environmental aspects associated with the activities at operating units should, where relevant, consider:

- emissions to air;
- releases to water bodies;
- waste management;
- contamination of land;
- use of raw materials and natural resources;
- other local environmental issues.

Significant EAs are the most important ones that cause the highest environmental impact or are important due to legislation and other requirements (e.g. environmental policy, customer demands). Significance equals the prioritizing (not relative) between chosen EAs at a company.

There are many EAs in the refining sector because the petroleum industry is among the highest generators of pollution. While it has made major strides to reduce emissions and hazardous wastes since about the late 1980s, it continues to generate significant levels of toxic air emissions and poorly manage many of its other EAs. There are numerous waste reduction case studies that have been documented where petroleum refineries have simultaneously reduced pollution and operating costs, but there are many more that are never implemented because a major barrier is cost. Because environmental accounting practices focus largely on direct financial returns, most pollution reduction options appear not to pay for themselves. Corporate mentality is such that investments typically must earn an adequate return on invested capital for the shareholders and some pollution prevention options at some facilities may not meet the requirements set by the companies. In addition, the equipment used in the petroleum refining industry is very capital intensive and has very long lifetimes. This reduces the incentive to make process modifications to (expensive) installed equipment that is still useful.

What is often missed is the fact that pollution prevention techniques are often more cost-effective than pollution reduction through end-of-pipe treatment. This is best understood when consideration is given to indirect cost savings. Indirect cost savings include: reduced healthcare costs from less exposure to air pollution; reduced threat from litigations for property damages, medical monitoring, and health claims by citizens who have been exposed to pollution; greater investor confidence (direct correlation between stock prices on the Dow Jones Industrials have been linked to reported reductions in emissions from company Toxic Release Inventory reporting). A further example is a case study based on the Amoco/Environmental Protection Agency (EPA) joint study, which claimed that the same pollution reduction currently realized through end-of-pipe regulatory requirements at the Amoco facility could be achieved at 15% of the current costs using pollution prevention techniques.

In addition to the general discounting of indirect cost savings, today's regulatory incentives to invest in cleaner production technologies are poor. Consider the following:

- The 1990 Clean Air Act Amendments intended to encourage voluntary reductions above the regulatory requirements by allowing facilities to obtain emission credits for voluntary emissions reductions. These credits were to serve as offsets against any potential future facility modifications resulting in an increase in emissions. Other regulations established by the amendments, however, require the construction of major new units within existing refineries to produce reformulated fuels. But these new operations require emission offsets in order to be permitted. This is counter-productive because it consumes many of the credits available for existing facility modifications. Thus a shortage of credits for facility modifications makes it difficult to receive credits for emission reductions through pollution prevention projects.

- Under the Clean Water Act, discharge of water-borne pollutants is limited by NPDES permits. Refineries that meet their permit requirements often have their permit limits changed to lower values. Because system upsets occur, resulting in significant excursions above the normal performance values, many refineries believe they must maintain a large operating margin below the permit limits in order to ensure continuous compliance. Refineries that can significantly reduce water-borne emissions are faced with the risk of having their permit limits lowered, which is a disincentive.

- Wastes failing a Toxicity Characteristic (TC) test are considered hazardous under the Resource Conservation and Recovery Act (RCRA). There is less incentive for a refinery to attempt to reduce the toxicity of such waste below the TC levels because, even though such toxicity reductions may render the waste non-hazardous, such waste may still have to comply with new land disposal treatment standards under subtitle C of the RCRA before being land disposed. There is less incentive to reduce the toxicity of listed refinery hazardous wastes because, once listed, the waste is subject to subtitle C regulations without regard to by how much the toxicity levels are reduced.

In addition to these disincentives, the USA has had 8 years of national policy that has heavily favored productivity at the expense of the environment and public safety. This is exemplified by attempts to dismantle the Toxic Release Inventory program, refusal to commit to international reduction targets in greenhouse gas emissions, the absence of an energy conservation policy, refusal to invest in renewable energy resources, and an extension of emissions monitoring requirements over longer periods with less frequent reporting.

While these factors suggest that the industry will continue to be satisfied with poor to marginal environmental performance, there has been an exponential increase in class action and private citizen suits against the industry and even the EPA. This in fact is simply history repeating itself. In a time before the US EPA it was citizen action groups and civil court actions on the part of citizens and class actions that forced corporations to act more responsibly.

This volume is written largely for the industry. It highlights EAs, offers alternatives to managing them with a focus on some of the more low-cost pollution prevention practices, and it is intended to stimulate ideas and approaches to better management of pollution issues.

There are six chapters. The first chapter provides an overview of gas plant operations and refineries, and identifies major EAs. Chapters 2–4 are case studies of major incidents that resulted in catastrophic releases of oil and refined products. Chapter 5 provides a critical assessment of the methodology and calculation procedures that the industry relies on in preparing emissions inventories. The chapter offers alternative approaches to providing more accurate emissions estimates. Chapter 6 provides guidelines on cleaner production and pollution prevention practices for improving overall environmental performance.

The authors wish to thank Elsevier for the fine production of this volume.

Nicholas P. Cheremisinoff Ph.D.
Paul Rosenfeld Ph.D.

1 The petroleum industry

1.1 Introduction

The petroleum industry refines crude petroleum and processes natural gas into a multitude of products. It is also involved in the distribution and marketing of petroleum-derived products. The primary family of pollutants emitted from these activities is volatile organic compounds (VOCs) arising from leakage, venting, and the evaporation of raw materials and finished products. The air emissions comprise point, fugitive, and area sources. Other significant air emissions include sulfur oxides, hydrogen sulfide, particulate matter, and a wide range of toxic chemicals. The operations within a typical refinery also emit a variety of criteria pollutants and toxic chemicals from fuel combustion devices. Oil- and gas-field operations as well as gas processing plants are also significant sources of emissions.

Historically the industry sector has not acted responsibly towards environmental management. Later chapters document poor environmental management practices that have stemmed from both unintentional and intentional actions. These actions have placed the public at risk from both chronic and acute exposures to various toxic chemicals, including significant amounts of carcinogens like benzene.

A problem with the sector is the lack of a systematic and transparent approach to the quantification and reporting of air emissions. The majority of air emissions from refinery operations are fugitive in nature. The literature that the authors have reviewed support that, on the whole, the industry continues to rely on the application of published emission factors that are not statistically significant and calculation procedures that favor low estimations. The under-reporting of air emissions has a significant advantage to companies because pollution fees imposed by regulators can be minimized and regulatory enforcement held in check. The sector thus has no direct financial incentives to improve on the accuracy of its quantification, reporting, and control of emissions.

In a report to Congress (Waxman report, 1999) it has been noted that oil refineries "vastly under-report leaks from valves to federal and state regulators and that these unreported fugitive emissions from oil refineries add millions of pounds of harmful pollutants to the atmosphere each year, including over 80 million pounds of volatile organic chemicals (VOCs) and over 15 million pounds of toxic pollutants."

Fugitive emissions are the emissions from equipment leaks, such as from valves, storage tanks, and various support equipment. Over 50% of all reported VOC and toxic air emissions from refineries are fugitive emissions according to the US Environmental Protection Agency (EPA). The Waxman report goes on to state that the "refineries fail to report large volumes of fugitive emissions. The average oil refinery reports ... that 1.3% of the valves at its facilities have leaks. In fact, the average leak rate from valves in refineries is 5.0% – nearly four times higher than the average reported leak rate." This under-reporting is alarming because it means that emissions reported under the Toxic Release Inventory (TRI) program in the USA are unreliable and cannot be used as a basis for assessing industry environmental performance and community risk.

As noted in the Preface, it is our intent to provide greater transparency to the identification and quantification of emissions and waste streams from refinery and gas processing operations. It is also the intent of this handbook to document best management practices, cleaner production technologies, and pollution prevention practices that can assist in improving environmental performance.

This first chapter provides an overview of the most widely used technologies employed by the industry. Many of the descriptions of refinery process operations are taken from the US OSHA standards and US EPA's AP-42 for background purposes. An identification of many of the sources of pollution is given along with these descriptions.

1.2   Oil- and gas-field operations

1.2.1   Field characterizations

Schlumberger World Energy Atlas lists more than 40,000 oil and gas fields of varying sizes throughout the world. Approximately 94% of known oil is concentrated in fewer than 1500 giant and major fields (Ivanhoe and Leckie, 1993). The largest discovered conventional oil field is the Ghawar Field in Saudi Arabia. Approximately 65% of all Saudi oil produced between 1948 and 2000 came from the Ghawar Field. Cumulative production to the end of 2005 was about 60 billion barrels (http://en.wikipedia.org/wiki/Ghawar_Field#cite_note-5). Currently it is estimated to produce over 5 million barrels (800,000 m^3) of oil a day, which is roughly equivalent to 6.25% of global production. Ghawar also produces approximately 2 billion cubic feet (57,000,000 m^3) of natural gas per day.

There are also massive unconventional oil fields, such as Venezuela's Orinoco tar sands and Canada's Athabasca tar sands. These fields reportedly may contain even greater reserves than the Ghawar Field.

Oil and gas fields are characterized by the geological structure of the field, as well as by the quality and composition of the production streams. Depending on the set of conditions, different and sometimes unique recovery processes are employed. Discoveries of new oil and gas reserves generally require drilling of very deep wells. As a consequence, the wellhead equipment must be capable of handling high-temperature/high-pressure hydrocarbons with a high degree of reliability.

Oil and gas reserves are brought to the surface through piping that runs the entire depth of the well, and which is hung within a steel casing. Since the casing diameter is larger than that of the piping, there is a void space or "annulus" between the tubing and the casing.

In many oil reservoirs the naturally occurring pressure is sufficient to force the crude oil to the surface of the well. This production process is referred to as "primary recovery" and most generally does not require the use of a compressor. However, the duration of the primary recovery is limited because at a certain point in time the natural energy to lift the oil is no longer adequate. After this point, a compressor and choke valve combination is used to restore or increase the pressure in the field. This phase of the well's life is known as *gas depletion* and is a form of secondary recovery.

In situations where the oil reservoir pressure is not sufficient to ensure the desired level of production, pumping systems may be employed. Enhanced recovery systems are installed to increase production and/or to avoid the decline of production over the years and to increase the recovery ratio.

1.2.2 Drilling rigs

Boreholes are made to recover oil and gas using a machine known as a drilling rig. They can be mobile equipment mounted on trucks, tracks or trailers, or more permanent land- or marine-based structures (such as oil platforms, commonly called "offshore oil rigs"). The term "rig" refers to the complex of equipment that is used to penetrate the surface of the Earth's crust. Small, portable systems are generally used for mineral exploration and drilling water wells, and in environmental investigations.

Larger, more fixed installations are capable of drilling through thousands of meters of the Earth's crust. Large "mud pumps" circulate drilling mud (slurry) through the drill bit and the casing, for cooling and removing the "cuttings" while a well is being drilled. Hoists in the rig can lift hundreds of tons of pipe. Other equipment can force acid or sand into reservoirs to facilitate extraction of the oil or mineral sample. Marine rigs may operate many hundreds of miles or kilometers offshore with infrequent crew rotation.

An example of an onshore rig is shown in Figure 1.1. The following list provides definitions of each of the equipment components shown in the diagram. The equipment associated with a rig depends on the type of rig, but typically includes at least some of the following items:

1. Mud tank – often called mud pits; provides a reserve store of drilling fluid until it is required down the wellbore.
2. Shale shakers – separate drill cuttings from the drilling fluid before it is pumped back down the wellbore.
3. Suction line (mud pump) – intake line for the mud pump to draw drilling fluid from the mud tanks.
4. Mud pump – reciprocal type of pump used to circulate drilling fluid through the system.

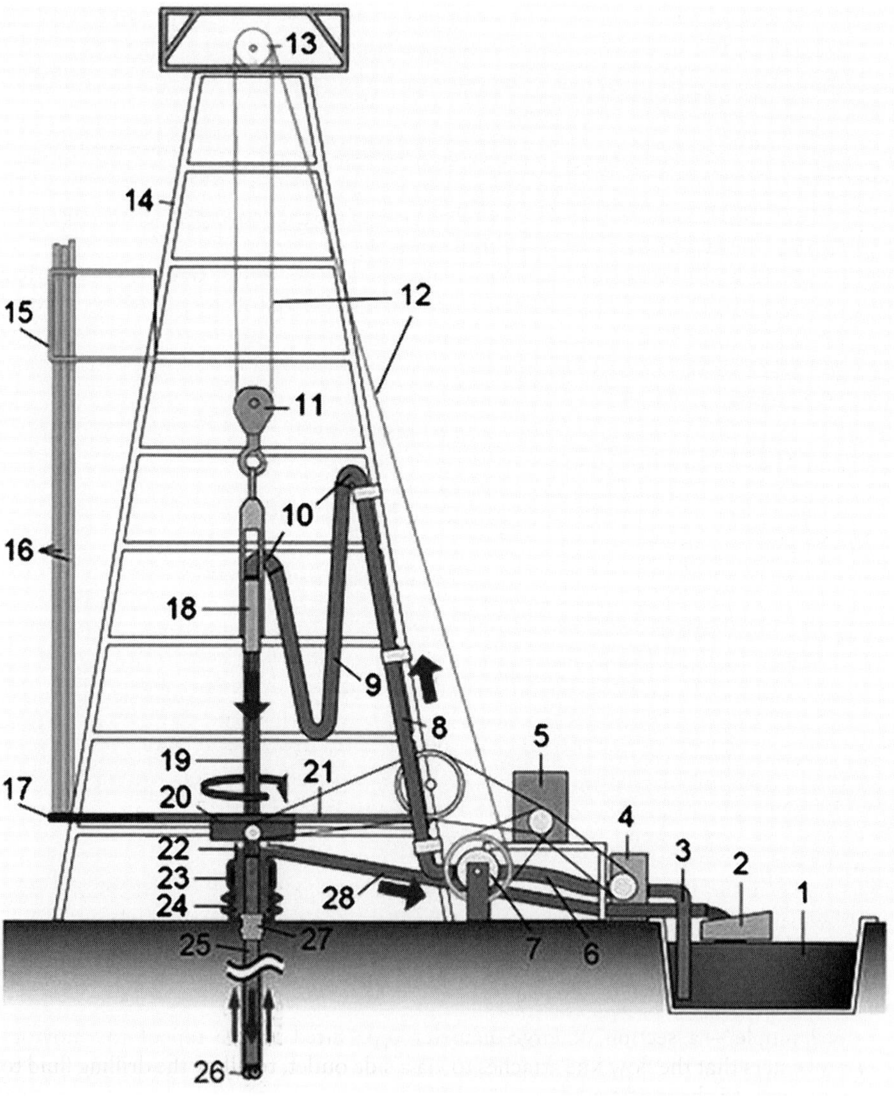

Figure 1.1 Details of a typical onshore drilling rig for oil recovery.
Diagram taken from Wikipedia: http://en.wikipedia.org/wiki/Drilling_rig

5. Motor or power source – a hydraulically powered device positioned just above the drill bit used to spin the bit independently from the rest of the drill string.

6. Vibrating hose – a flexible, high-pressure hose (similar to the *kelly hose*) that connects the mud pump to the *standpipe*. It is called the *vibrating hose* because it tends to vibrate and shake (sometimes violently) due to its close proximity to the mud pumps.

7. Draw-works – the mechanical section that contains the spool, whose main function is to reel in/out the drill line to raise/lower the traveling block.

8. Standpipe – a thick metal tubing, situated vertically along the derrick, that facilitates the flow of drilling fluid and has attached to it and supports one end of the kelly hose.

9. Kelly hose – a flexible, high-pressure hose that connects the standpipe to the *kelly* (or more specifically to the gooseneck on the swivel above the kelly) and allows free vertical movement of the kelly, while facilitating the flow of the drilling fluid through the system and down the drill string.

10. Goose-neck – thick metal elbows connected to the swivel and standpipe that supports the weight of and provides a downward angle for the kelly hose to hang from.

11. Traveling block – moving end of the *block and tackle*; together they give a significant mechanical advantage for lifting.

12. Drill line – thick, stranded metal cable threaded through the two blocks (traveling and crown) to raise and lower the *drill sting*.

13. Crown block – stationary end of the *block and tackle*.

14. Derrick – the support structure for the equipment used to lower and raise the drill string into and out of the wellbore.

15. Monkey board – the structure used to support the top end of the stands of drill pipe vertically situated in the derrick.

16. Stand (of drill pipe) – sections of two or three joints of drill pipe connected together and *stood* upright in the derrick. When pulling out of the hole, instead of laying down each joint of drill pipe, two or three joints are left connected together and stood in the derrick to save time.

17. Pipe rack (floor) – a part of the drill floor (#21) where the stands of drill pipe are stood upright, typically made of a metal frame structure with large wooden beams situated within it. The wood helps to protect the end of the drill pipe from damage.

18. Swivel (on newer rigs this may be replaced by a top drive).

19. Kelly drive – drive unit.

20. Rotary table – rotates, along with its constituent parts the kelly and kelly bushing, the drill string and the attached tools and bit.

21. Drill floor – the area on the rig where the tools are located to make the connections of the drill pipe, bottom hole assembly, tools and bit. It is considered the main area where work is performed.

22. Bell nipple – a section of large-diameter pipe fitted to the top of the blowout preventers that the flow line attaches to via a side outlet, to allow the drilling fluid to flow back to the mud tanks.

23. Blowout preventer (BOP) annular – annular (often referred to as the *Hydril*, which is one manufacturer) and pipe rams and blind rams (see #24).

24. Blowout preventers (BOPs) pipe ram and blind ram – devices installed at the wellhead to prevent fluids and gases from unintentionally escaping from the wellbore.

25. Drill string – an assembled collection of drill pipe, heavyweight drill pipe, drill collars, and any of a whole assortment of tools, connected together and run into the wellbore to facilitate the drilling of a well, the collection of which is referred to singularly as the drill string.

26. Drill bit – device attached to the end of the drill string that breaks apart the rock being drilled. It contains jets through which the drilling fluid exits.

27. Casing head – a large metal flange welded or screwed onto the top of the conductor pipe (also known as *drive pipe*) or the casing and used to bolt the surface equipment to equipment such as the blowout preventers (for well drilling) or the Christmas tree (for well production).
28. Flow line – large-diameter pipe that is attached to the bell nipple and extends to the shale shakers to facilitate the flow of drilling fluid back to the mud tanks.

The photographs shown in Figures 1.2–1.6 provide examples of smaller drilling rigs. The reader's attention is drawn to the various valves, connectors, flanges, the wellhead, and different joints among the piping aperture. Each of these entities is a source of fugitive emissions and leaks. There are multiple emission points with some of the components.

Valves, flanges, pumps, connectors, compressors, and drains are each sources for leaks and fugitive emissions. The method adopted in the USA for estimating fugitive emissions from these sources was developed for the industry sector by the US Environmental Protection Agency (EPA) in 1995 and the American Petroleum Institute (API) in 1996. The following data are required for average emission factors estimation calculations:

- the numbers of each type of component (e.g. valves, flanges, etc.) in each process unit;
- the service each component is in (e.g. gas, light liquid, heavy liquid, water/oil). The definitions in Table 1.1 are applied when determining the type of "service" a particular piece of equipment is in, so that the appropriate emission factors are used;
- the weight fraction of total organic compounds (TOCs) within the stream;
- hours operational (e.g. hours/year). Hours of operation should be determined for particular streams.

Figures 1.7–1.9 provide drawings that identify the locations among different piping components where leaks and fugitive emissions can occur.

According to the US EPA (Hummel, 1990) the average gas wellhead component count consists of 11 valves, 50 screwed connections, one flange, and two open-ended lines. By way of example, 17 new production wells introduce 1088 more piping components to an existing field, all of which are potentially new sources of leaks and fugitive emissions.

Emissions from oil and gas wellheads can be estimated using the average emission factor approach as indicated in the EPA *Protocol for Equipment Leak Emission Estimates* (US EPA, 1995). This method uses average emission factors in combination with wellhead-specific data. These data include: (1) number of each type of component (valves, flanges, etc.); (2) the service type of each component (gas, condensate, mixture, etc.); (3) the benzene concentration of the stream; and (4) the number of wells. A main source of data for equipment leak hydrocarbon emission factors for oil and gas field operations is an API study (Eaton et al., 1980).

Benzene and total hydrocarbons equipment leak emission factors from oil wellheads are reported in Table 1.2 (obtained from Serne et al., 1991). These emission factors were developed from screening and bagging data obtained in oil

Figure 1.2 Operating drilling rigs for gas recovery wells.

Figure 1.3 Wellhead.

production facilities located in California, as reported by Serne et al. (1991). In this study more than 450 accessible production wellhead assemblies were screened, and a total of 28 wellhead assemblies were selected for bagging. The emission factors reported in Table 1.2 are for field wellheads only. The factors do not include other field equipment such as dehydrators, separators, in-line heaters, treaters, or any other equipment.

Figure 1.4 Connectors and valves at well drilling rig.

Figure 1.5 Flanges, connectors, and valves at a drilling rig.

Figure 1.6 Piping components at well drilling rig.

Table 1.1 Standard API definitions of materials handled by different types of equipment

Service type	Standard definition
Gas/vapor	Product is in gas state at operating conditions
Light liquid	Product is in liquid state in which the sum of the concentrations of individual constituents with a vapor pressure over 0.3 kPa at 20°C is greater than or equal to 20 wt%
Heavy liquid	Product does not fall under the classifications for gas/vapor, or light liquid service
Water/oil	Water streams in oil service with a water content greater than 50% from the point of origin to the point where the water content reached 99%. For water content streams with a water content greater than 99% the emissions rate is considered negligible

Source: API (1996).

The composition of gas streams varies among production sites. Therefore, when developing benzene emission estimates, the total hydrocarbons emission factors need to be modified by specific benzene weight percentage, if such data are available. Benzene constituted from less than 0.1% up to 2.3% weight

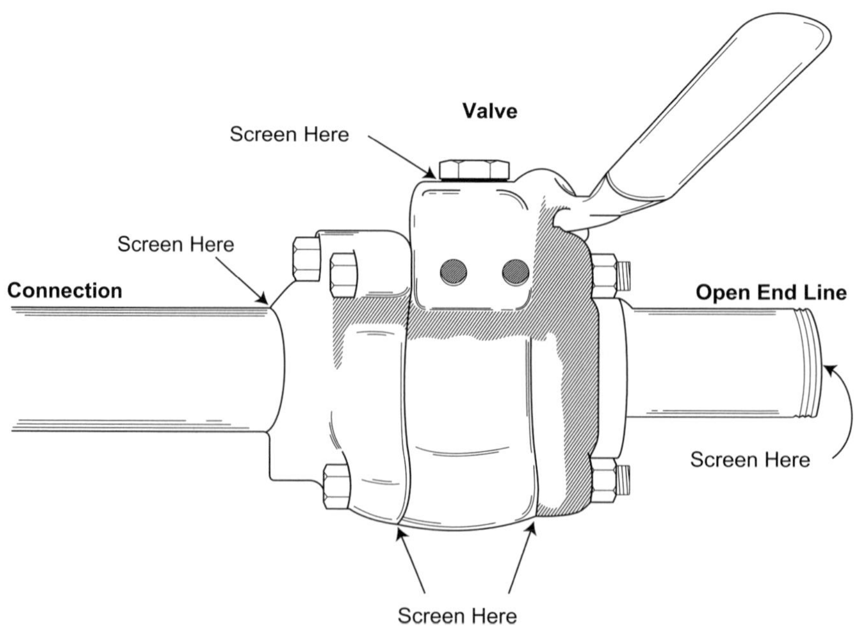

Figure 1.7 Drawing of a valve showing screening locations for leaks.

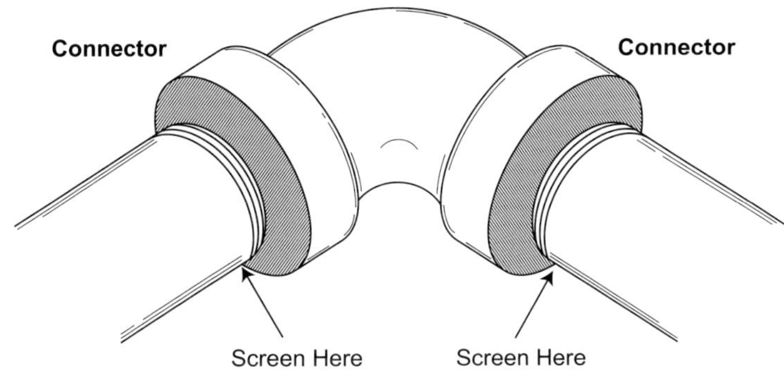

Figure 1.8 Drawing of a connector showing screening locations for leaks.

of total non-methane hydrocarbons (TNMHCs) for water flood wellhead samples from old crude oil production sites in Oklahoma. The literature also notes that VOC composition in the gas stream from old production sites is different than that from a new field. It is therefore good practice for the operator to perform an analysis of the gas composition beforehand to verify use of emission factors.

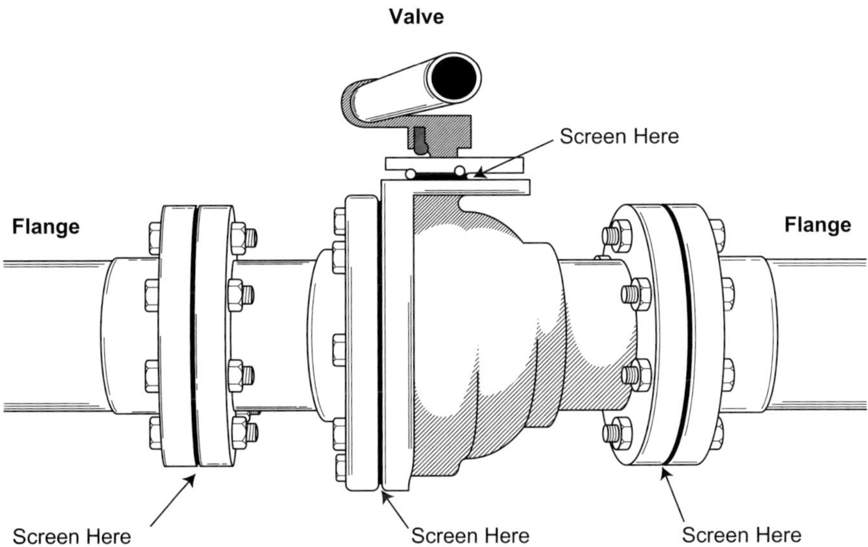

Figure 1.9 Drawing of a flanged valve showing screening locations for leaks.

Table 1.2 Benzene and total hydrocarbons equipment leak emissions factors for oil wellhead assemblies

Emission level[a]	Total hydrocarbons		Benzene		Emission factor rating
	lb/h per wellhead	kg/h per wellhead	lb/h per wellhead	kg/h per wellhead	
1	3.67E−02	1.65E−02	1.27E−07	5.77E−08	D
2	6.53E−03	2.97E−03	3.90E−08	1.77E−08	D
3	9.74E−04	4.43E−04	6.25E−09	2.84E−09	D
4	3.48E−04	1.58E−04	NA	NA	D
5	1.06E−04	4.82E−05	NA	NA	D

[a]The concentration ranges applicable to the five emission levels developed were as follows. Level 1: >10,000 ppm at two or more screening points or causing instrument flameout; Level 2: 3000–10,000 ppm; Level 3: 500–3000 ppm; Level 4: 50–500 ppm; Level 5: 0–50 ppm.
Source: Serne et al. (1991).

1.2.3 Reinjection

Reinjection is a method of enhanced oil recovery used to compensate for the natural decline of an oil field's production by increasing the pressure in the reservoir, thus restoring the desired level of production and stimulating the recovery of additional crude oil. Using this technique the field exploitation can be increased by as much as 20%. Gas that is reinjected is usually the associated gas separated from the crude oil in the flash and stabilization phases. Other gases, such as nitrogen or carbon dioxide, may also be used for this purpose. The gas is reinjected into the reservoir in dedicated wells and forces the oil to migrate toward the well bores of the producing wells. More recent material technology advances allow associated sour gases containing high percentages of H_2S and/or CO_2 to be reinjected without the need for sweetening.

Depending on the depth and physical characteristics of the field, high injection pressures may be required. High-pressure barrel compressors are used in this application. In the case of moderate gas flows reciprocating compressors may be employed. As an example, GE Oil & Gas reported that its pressure centrifugal compressor was operated at 820 bar discharge pressure handling an extremely sour gas with 18% H_2S.[1]

Water injection is another technique used for enhanced oil recovery. The production water separated from the crude oil is filtered, treated, and then reinjected into the reservoir by means of high-pressure centrifugal pumps.

1.2.4 Boosting stations

Natural gas is moved from producing regions to points of consumption by means of a network of pipelines. Gas boosting stations are installed along

[1] http://www.gepower.com/businesses/ge_oilandgas/en/applications/reinject_plant.htm

a pipeline at variable distances (100–250 km) to compensate for the loss in gas pressure that occurs along the pipeline to ensure the adequate flow of product at the delivery end of the pipeline. Pipelines with large diameters transport large amounts of gas over long distances (interstate or cross-regional) at pressures ranging from 60 to 100 bar. Small/medium pipelines distribute smaller amounts of gas to usage points across shorter distances and at lower pressures. Boosting stations are employed to recover lost pumping energy along the transmission lines.

A boosting station includes inlet separation facilities to remove liquid droplets that may be present, a compression unit to increase the gas pressure, a cooling system, and station auxiliaries and a control system. The heart of the station is the compression unit, which is generally a gas turbine-driven centrifugal compressor having a low compression ratio and large gas flow. The gas turbine uses a small portion of the same gas that is compressed as its fuel. Therefore the combined efficiency of the turbo-compressor unit is of importance.

Electric motor-driven compressors are often used in populated or highly industrialized areas to cope with emission restrictions. This practice does not consume any of the compressed gas; however, it requires a reliable source of electrical power.

Reciprocating compressors driven either by reciprocating gas engines or by electric motors may also be used in small pipeline applications.

Oil pipelines are used to transport oil from the production area to the export loading terminal, or to a processing unit, such as a refinery. Oil boosting stations are installed along the pipeline at variable distances, to compensate for the pipeline pressure losses and to ensure a constant flow of oil. In an oil boosting station, one or more high-capacity, single or multistage centrifugal pumps are employed. These are often driven by a gas turbine, or diesel engine, or by an electric motor up to a unit power range of about 30 MW.

1.3 Gas plant processing operations

The natural gas product fed into the mainline gas transportation system must meet specific quality measures in order for the pipeline grid to operate properly. Consequently, natural gas produced at the wellhead, which contains contaminants and natural gas liquids, must be processed, i.e. cleaned, before it can be safely delivered to the high-pressure, long-distance pipelines that transport the product to the consumer. Natural gas that is not within certain specific gravities, pressures, Btu content range, or water content levels will cause operational problems, pipeline deterioration, or pipeline rupture. The contaminants found in natural gas at the wellhead include non-hydrocarbon gases such as water vapor, carbon dioxide, hydrogen sulfide, nitrogen, oxygen, and helium. Ethane, propane, and butane are the primary heavy hydrocarbons (liquids) extracted at a natural gas processing plant, but other petroleum gases, such as isobutane, pentanes, and normal gasoline, also may be processed. Figure 1.10 provides a generalized schematic of a typical gas processing plant.

The natural gas received and transported by the interstate mainline transmission systems must meet the quality standards specified by pipeline companies in the "General Terms and Conditions (GTC)" section of their tariffs. These quality standards vary from pipeline to pipeline and are usually a function of a pipeline system's design, its downstream interconnecting pipelines, and its customer base. In general terms, these standards specify that the natural gas:

- be within a specific Btu content range (1035 Btu per cubic feet, ± 50 Btu);
- be delivered at a specified hydrocarbon dew point temperature level (below which any vaporized gas liquid in the mix will tend to condense at pipeline pressure);
- contain no more than trace amounts of elements such as hydrogen sulfide, carbon dioxide, nitrogen, water vapor, and oxygen;
- be free of particulate solids and liquid water that could be detrimental to the pipeline or its ancillary operating equipment.

Gas processing equipment is intended to assure that these tariff requirements can be met. In recent years, as natural gas pricing has transitioned from a volume basis (per thousand cubic feet) to a heat-content basis (per million Btu), producers have tended, for economic reasons, to increase the Btu content of the gas delivered into the pipeline grid while decreasing the amount of natural gas liquids extracted from the natural gas stream. Consequently, interstate pipeline companies have had to monitor and enforce their hydrocarbon dew point temperature level restrictions more frequently in order to avoid potential liquid formation within the pipes that may occur as a result of producers maximizing the Btu content.

This has implications regarding increased emissions for gas processing plants with limited customer bases. Because pipeline customers may reject the gas because of poor quality, the customer may literally shut the valve, resulting in stoppages at the gas plant. This action causes pressure to build up and if another customer such as a refinery is not available to take the spare capacity, gas will have to be flared. Flaring operations will result in increased emissions from the gas processing plant. In most facilities that sell the gas, a gas transmission pipeline will have the gas composition monitored continuously with a gas chromatograph. Another reason that may cause the sell valve to close is customer equipment failures. This further places the gas processing plant at a disadvantage and there is a risk of increased emissions from unplanned stoppages.

As already noted, natural gas processing begins at the wellhead. The composition of the raw natural gas extracted from producing wells depends on the type, depth, and location of the underground deposit and the geology of the area. Oil and natural gas are found together in the same reservoir. The natural gas produced from oil wells is generally classified as "associated-dissolved", meaning that the natural gas is associated with or dissolved in crude oil. Natural

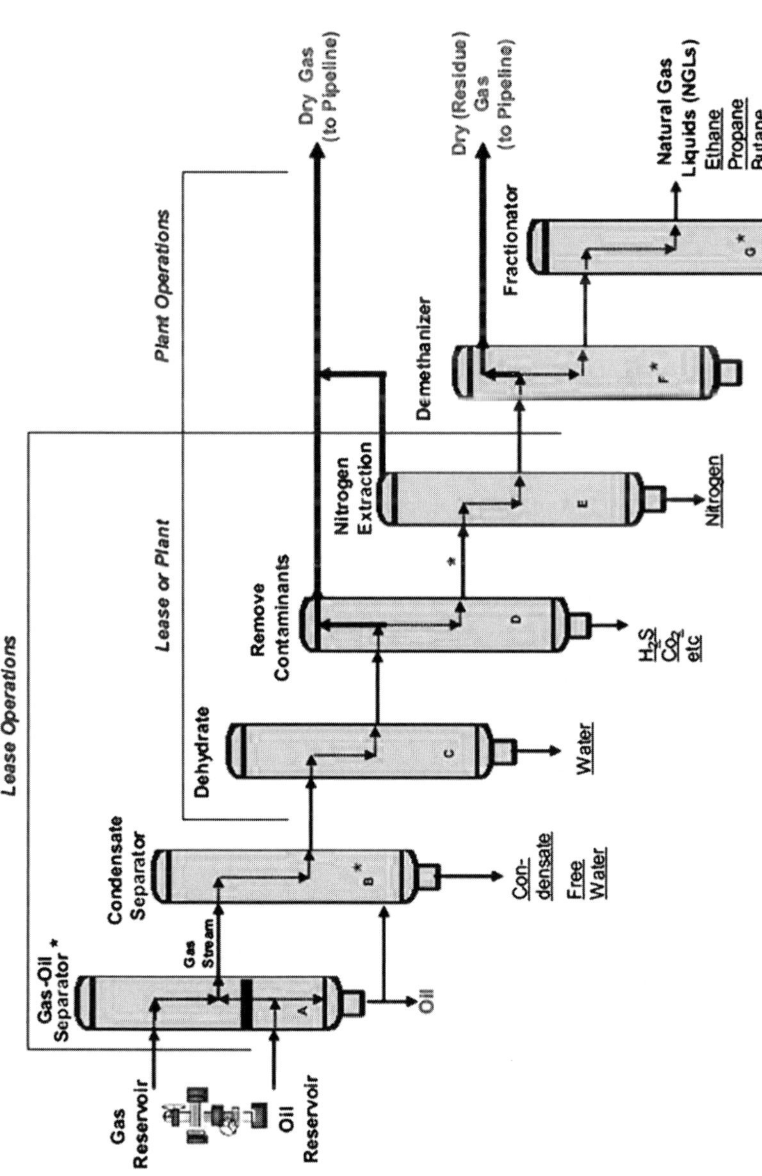

Figure 1.10 Generalized schematic of a natural gas processing plant.
Source: Energy Information Administration, Office of Oil and Gas, January 2006.

gas production absent from any association with crude oil is classified as "non-associated". About 75% of US wellhead production of natural gas is non-associated.

Natural gas production contains small (two to eight carbons) hydrocarbon molecules in addition to methane. Although they exist in a gaseous state at underground pressures, these molecules become liquid (i.e. the gas condenses) at normal atmospheric pressure. Collectively, they are referred to as condensates or natural gas liquids (NGLs).

The processing of wellhead natural gas into pipeline-quality dry natural gas involves several processes to remove oil, water, elements such as sulfur, helium, and carbon dioxide, and natural gas liquids. In addition to those four, it is necessary to install scrubbers and heaters at or near the wellhead. The scrubbers serve primarily to remove sand and other large-particle impurities, and they are a source of air emissions. The heaters ensure that the temperature of the natural gas does not drop too low and form a hydrate with the water vapor content of the gas stream. Natural gas hydrates are crystalline ice-like solids or semi-solids that can impede the passage of natural gas through valves and pipes.

The wells from an oil and gas field are connected to downstream facilities via a process called gathering, wherein small-diameter pipes connect the wells to initial processing/treating facilities. Beyond the fact that a producing area can occupy many square miles and involve hundreds of wells, each with its own production characteristics, there may be a need for intermediate compression, heating, and scrubbing facilities, as well as treatment plants to remove carbon dioxide and sulfur compounds, prior to the processing plant. All of these factors make gathering-system design complex, and further complicate quantification of emissions.

Non-pipeline-quality gas is generally piped to a central natural gas processing plant for liquids extraction and eventual delivery of pipeline-quality natural gas at the plant tailgate. The gas processing plant receives gas from a gathering system and sends out the processed gas via an output (tailgate) lateral that is interconnected to major interstate pipeline networks. Liquids removed at the processing plant are taken away by pipeline to petrochemical plants, refineries, and other gas liquids customers. Some of the heavier liquids are temporarily stored in tanks on-site and then transported in tankers to customers.

The following are definitions of the unit process technologies that are used in a typical natural gas processing plant:

- **Gas–oil separation.** Usually pressure relief at the wellhead will result in a natural separation of gas from oil (using a conventional closed tank, where gravity separates the gas hydrocarbons from the heavier oil). In some cases, however, a multistage gas–oil separation process is needed to separate the gas stream from the crude oil. Gas–oil separators are usually closed cylindrical shells, horizontally mounted with inlets at one end, an outlet at the top for removal of gas, and an outlet at the bottom for removal of oil. Separation is accomplished by alternately heating and cooling (by compression) the flow stream through multiple steps. Water and condensate are also extracted by this process.

- **Condensate separation.** Condensates are removed from the gas stream at the wellhead through the use of mechanical separators. The more common arrangement is to first send the gas to separators. After separation the gas stream enters the processing plant at high pressure (usually 600 psig or greater) through an inlet slug catcher, where free water is removed from the gas, after which it is directed to a condensate separator. Extracted condensate can be routed to on-site storage tanks.
- **Dehydration.** A dehydration process is employed to remove water, which may cause the formation of hydrates. Hydrates form when a gas or liquid containing free water experiences specific temperature/pressure conditions. Dehydration is defined as the removal of water from the produced natural gas and can be accomplished by several technologies. A common system is an ethyleneglycol (glycol injection) system, which serves as an absorption mechanism to remove water and other solids from the gas stream. An alternative technology is adsorption dehydration, which uses dry-bed dehydrator towers. These employ desiccants such as silica gel and activated alumina to accomplish the extraction.
- **Contaminant removal.** This unit process is designed to remove hydrogen sulfide, carbon dioxide, water vapor, helium, and oxygen. In this process the gas flows though a tower containing an amine solution. Amines absorb sulfur compounds from natural gas and can be reused repeatedly. After desulfurization, the gas flow is directed to the next section, which contains a series of filter tubes. As the velocity of the stream reduces in the unit, primary separation of remaining contaminants occurs via gravity. Separation of smaller particles occurs as gas flows through the tubes, where they combine into larger particles that flow to the lower section of the unit. As the gas stream continues through the series of tubes, a centrifugal force is generated, which further removes additional water and small solid particulates.
- **Nitrogen extraction.** In some natural gas processing plants, once the hydrogen sulfide and carbon dioxide are processed, the stream is routed to a nitrogen rejection unit (NRU), where it is further dehydrated using molecular sieve beds. In the NRU, the gas stream is routed through a series of passes through a column and a brazed aluminum plate fin heat exchanger. The nitrogen is cryogenically separated and vented. The venting is a source of emissions. Another type of NRU separates methane and heavier hydrocarbons from nitrogen using an absorbent solvent. The absorbed methane and heavier hydrocarbons are flashed off from the solvent by reducing the pressure of the processing stream in multiple gas decompression steps. The liquid from the flash regeneration step is returned to the top of the methane absorber as a lean solvent. Helium can be extracted from the gas stream through membrane diffusion in a pressure swing adsorption (PSA) unit.
- **Methane separation.** The industry generally employs cryogenic processing and absorption methods as the means to separate methane from NGLs. Cryogenic processing consists of lowering the temperature of the gas stream (typically -120 to $-30°F$). The authors believe this is accomplished with a turbo expander process using external refrigerants to chill the gas stream. There is a rapid drop in temperature, which condenses the hydrocarbons in the gas stream but allows methane to remain in its gaseous form.
- **Fractionation.** Fractionation is the process of separating the various NGLs present in the remaining gas stream. As in the case of a refinery, fractionation uses the varying boiling points of the individual hydrocarbons in the stream, which are NGLs in this step of the process, to achieve the separation. The process occurs in stages as the gas

stream rises through a tower where heating units raise the temperature of the stream, causing the various liquids to separate and exit into specific holding tanks. Propane can be separated out and sent to storage and then sold. Butane and heavier hydrocarbons can also be separated. These can be blended in crude oil and sent off to a separate pipeline for sales.

In general, the fugitive emissions from oil and gas activities are from the following primary sources:

- fugitive equipment leaks;
- process venting;
- evaporation losses;
- disposal of waste gas streams (e.g. by venting or flaring);
- accidents and equipment failures.

Accident and equipment failures may include:

- well blowouts;
- pipeline breaks;
- tanker accidents;
- tank explosions
  - gas migration to the surface around the outside of wells (e.g. leak in the production string at some point below the casing surface, or migration of material from one or more of the hydrocarbon-bearing zones that are penetrated);
  - surface-casing vent blows (e.g. leak from the production casing into the surface casing, or fluid migration up into the surface casing from below).

Additional sources of fugitive emissions include:

- leakage of chlorofluorocarbons (CFCs) from refrigeration systems and SF_6 from electrical components;
- land disposal of solid waste;
- hydrocarbon emissions from wastewater handling;
- hydrocarbon emissions from industrial wastewater and sludge streams.

Each of the above is a source of emissions that cumulatively may contribute to significant air releases. Accurate assessment of each contribution is not straightforward, not only because of the significant numbers of sources, but because there are several factors affecting fugitive emissions and the speciation profile of the emissions. Among the more important factors are:

- the amount and type of infrastructure employed;
- the amount of waste-gas created;
- the incentives and requirements to control waste-gas volumes and reduce fugitive emissions.

These in turn are a function of the following:

- design and operating practices;
- frequency of maintenance and inspection activities;

- type, age, and quality of equipment;
- type of hydrocarbons being produced or handled and their composition;
- operating conditions;
- throughputs;
- pumping and compression requirements;
- metering requirements;
- treatment and processing requirements;
- frequency and duration of process upsets;
- sweet, sour, or odorized service;
- population density near the facility (i.e. proximity of sensitive receptors);
- applicable environmental and conservation regulations.

Current inventory methodologies available to the oil and gas industry are:

- Tier 1: Top-down average emission factor approach;
- Tier 2: Mass balance approach;
- Tier 3: Rigorous bottom-up approach.

While emission factors are used to develop and quantify an emissions inventory, the methodology employed by many facilities is generally unreliable and can lead to a significant understatement of total facility emissions. This criticism is substantiated by the discussions presented in Chapter 3.

1.4 Refining and refinery operations

1.4.1 Overview

The US petroleum industry began with the drilling of the first commercial oil well in 1859, followed by the opening of the first refinery 2 years later to process the crude into kerosene. Petroleum refining has evolved in response to changing consumer demand for better and different products. The original intent was to produce kerosene as a cheaper and improved alternate fuel for light over whale oil. The development of the internal combustion engine led to the production of gasoline and diesel fuels. The evolution of the airplane created a need first for high-octane aviation gasoline and then for jet fuel, which is a sophisticated form of the original product, kerosene.

Modern-day refineries produce a variety of products including many required as feedstock for the petrochemical industry. The evolution of petroleum refining from simple distillation to today's complex processes has been one of remarkable technological achievements, but it has brought along devastating effects on the environment and communities that live near and work within the industry. This in part is due to an aging infrastructure. The costs for constructing new refineries is almost cost prohibitive, especially in light of environmental restrictions in most countries. The last refinery built in the USA was in Garyville, Louisiana, which started up in 1976. Industry officials estimate the cost of building a new refinery at between $2 billion and $4 billion. The approval for constructing and operating a new refinery requires collecting up to 800 different permits (source: *Investor's Business Daily*). Aging infractructure means that the industry relies on old technologies, which in turn translates into inefficiency, waste, and pollution. This aging infrastructure means that communities living near refinery operations face potential disasters from exposure to pollution.

1.4.2 Crude oils and compositions

Hydrocarbons

Most commercial descriptions of the types of oil center around density, e.g. light crude, heavy crude, etc. API gravity, the American Petroleum Institute's measure of oil density, is the industry's most used standard.

The US National Bureau of Standards established the Baumé scale (degrees Baumé) as the standard for measuring specific gravity, or density of liquids less dense than water. API gravity is a measure of how heavy or light a petroleum liquid is compared to water.

The formula for calculating API gravity is:

$$\text{API gravity} = (141.50/\text{Specific gravity}) - 131.50$$

In general 40–45 API gravity degree oils have the greatest commercial value because they are rich in gasoline. Condensates are worth slightly less because the natural gasoline has a lower octane value. Heavier crudes are worth less because they require more refinery processing. West Texas Intermediate (WTI) is the benchmark crude oil used by the USA to set prices and compare other oils. It has 38–40 API gravity.

Crude oil containing free sulfur, hydrogen sulfide (H_2S), or other sulfur-containing compounds in amounts greater than 1% is considered sour crude. As is the case with sour gas, the sulfurs must be removed from the crude oil before the oil can be refined and the refiner pays less for oil that contains sulfur. Sour crude is usually processed into heavy oil such as diesel and fuel oil rather than gasoline to reduce processing costs. Sweet crude is oil that contains little or no sulfur.

All crudes contain both paraffinic and naphthenic components but are differentiated based on the level of those components. Paraffinic is crude oil containing a relatively high percentage (by volume) of linear and branched paraffins. Most conventional engine lubricating oils today are made from paraffinic crude oil. As the name suggests, paraffinic crude contains wax and generally has a higher API gravity, i.e. it is a lighter crude. The term paraffinic is often assumed to be synonymous with wax. In lubricating oils, the wax is removed by a refining process called dewaxing.

Naphthenic (asphaltic) crude contains relatively little wax. Naphthenic crude oils contain mainly (by volume) naphthenes and other aromatic hydrocarbons. They generally have a lower API gravity, i.e. they are the heavier crudes. They also contain other materials including nickel, iron, vanadium, and arsenic.

Crude oils are complex mixtures containing different hydrocarbon compounds that vary in appearance and composition from one oil field to another. Crude oils range in consistency from watery mixtures to tar-like solids. They also range in color from clear to black.

An "average" crude oil contains about 84% carbon, 14% hydrogen, 1–3% sulfur, and less than 1% each of nitrogen, oxygen, metals, and salts. As already noted, crude oils are generally classified as:

- paraffinic
- naphthenic, or
- aromatic.

Classification is based on the predominant proportion of similar hydrocarbon molecules. Mixed-base crudes have varying amounts of each type of hydrocarbon. Refinery crude base stocks usually consist of mixtures of two or more different crude oils.

Crude oil assays are used to classify crude oils as paraffinic, naphthenic, aromatic, or mixed. One assay method (United States Bureau of Mines) is based on distillation, and another method (UOP "K" factor) is based on gravity and boiling points. There are more comprehensive crude assays that can be used to determine the value of the crude (i.e. its yield and quality of useful products) and processing parameters. Crude oils are grouped according to yield structure.

We also define crude oils in terms of API gravity: the higher the API gravity, the lighter the crude. For example, light crude oils have high API gravities and low specific gravities. Crude oils with low carbon, high hydrogen, and high API gravity are rich in paraffins and tend to yield greater proportions of gasoline and light petroleum products, whereas those with high carbon, low hydrogen, and low API gravities are usually rich in aromatics.

The term "sour" is used to describe crude oils that contain appreciable quantities of hydrogen sulfide or other reactive sulfur compounds. In contrast, those crude oils with less sulfur are called "sweet". We have noted exceptions to this rule – one being West Texas crudes, which are always considered "sour" regardless of their H_2S content; another is Arabian high-sulfur crudes, which are not considered "sour" because their sulfur compounds are not highly reactive. Table 1.3 reports average literature values of the properties of types of crude and their gasoline potential. The values reported in Table 1.3 are taken from various reported sources on the Web and are averages of reported values.

Crude oil is a mixture of hydrocarbon molecules, which are organic compounds of carbon and hydrogen atoms that may include from one to 60 carbon atoms. The properties of the hydrocarbons depend on the number and arrangement of the carbon and hydrogen atoms in the molecules. The simplest hydrocarbon molecule is one carbon atom linked with four hydrogen atoms: methane. All other variations of petroleum hydrocarbons evolve from this molecule. Hydrocarbons containing up to four carbon atoms are usually gases, those with five to 19 carbon atoms are usually liquids, and those with 20 or more are solids.

Table 1.3 Average properties of different crudes and their gasoline potentials

Crude source	Paraffins (% vol)	Aromatics (% vol)	Naphthene (% vol)	Sulfur (% wt)	API gravity (approx.)	Naphth. yield (% vol)	Octane no. (typical)
Nigerian – Light	37	9	54	0.2	36	28	60
Saudi – Light	63	19	18	2.0	34	22	40
Saudi – Heavy	60	15	25	2.1	28	23	35
Venezuela – Heavy	35	12	53	2.3	30	2	60
Venezuela – Light	52	14	34	1.5	24	18	50
USA – Mid cont. sweet	–	–	–	0.4	40	–	–
USA – W. Texas sour	46	22	32	1.9	32	33	55
North Sea – Brent	50	16	34	0.4	37	31	50

The refining process uses chemicals, catalysts, heat, and pressure to separate and combine the basic types of hydrocarbon molecules naturally found in crude oil into groups of similar molecules. The refining process also rearranges the structures and bonding patterns into different hydrocarbon molecules and compounds. It is the type of hydrocarbon (paraffinic, naphthenic, or aromatic) rather than its specific chemical compounds that is significant in the refining process.

The paraffinic series of hydrocarbon compounds have the general formula C_nH_{2n+2} and can have either straight chains (normal) or branched chains (isomers) of carbon atoms. The lighter, straight-chain paraffin molecules are found in gases and paraffin waxes. Examples of straight-chain molecules are methane, ethane, propane, and butane (gases containing from one to four carbon atoms), and pentane and hexane (liquids with five to six carbon atoms). The branched-chain (isomer) paraffins are found in heavier fractions of crude oil and have higher octane numbers than normal paraffins. These compounds are saturated hydrocarbons, with all carbon bonds filled.

Aromatics are unsaturated ring-type (cyclic) compounds that react readily because they have carbon atoms that are deficient in hydrogen. All aromatics have at least one benzene ring as part of their molecular structure. Naphthalenes are fused double-ring aromatic compounds. The most complex aromatics, polynuclears (three or more fused aromatic rings), are found in heavier fractions of crude oil.

Naphthenes are saturated hydrocarbon groupings with the general formula C_nH_{2n}, arranged in the form of closed rings (cyclic) and found in all fractions of crude oil except the very lightest of crudes. Single-ring naphthenes (mono-cycloparaffins) with five and six carbon atoms predominate, with two-ring naphthenes (dicycloparaffins) found in the heavier ends of naphtha.

Other important hydrocarbons are alkenes, diolefins, and alkynes. Alkenes are mono-olefins with the general formula C_nH_{2n} and contain at least one carbon–carbon double bond in the chain. The simplest alkene is ethylene, with two carbon atoms joined by a double bond and four hydrogen atoms. Olefins are usually formed by thermal and catalytic cracking and rarely occur naturally in unprocessed crude oil.

Dienes, also known as diolefins, have two carbon–carbon double bonds. The alkynes are another class of unsaturated hydrocarbons. These have a carbon–carbon triple bond within the molecule. Both these series of hydrocarbons have the general formula C_nH_{2n-2}. Diolefins such as 1,2-butadiene and 1,3-butadiene, and alkynes such as acetylene, occur in C_5 and lighter fractions from cracking. The olefins, diolefins, and alkynes are unsaturated because they contain less than the amount of hydrogen necessary to saturate all the valences of the carbon atoms. These compounds are more reactive than paraffins or naphthenes and readily combine with other elements such as hydrogen, chlorine, and bromine.

Non-hydrocarbons

Other chemicals found in crude oils include sulfur compounds, oxygen compounds, nitrogen compounds, and trace metals.

Sulfur is generally present in crude oil as hydrogen sulfide (H_2S), as compounds (mercaptans, sulfides, disulfides, thiophenes, etc.) or as elemental sulfur. Each crude oil has different amounts and types of sulfur compounds, but as a rule the proportion, stability, and complexity of the compounds are greater in heavier crude-oil fractions. Hydrogen sulfide is a primary contributor to corrosion in refinery processing units. Other corrosive substances are elemental sulfur and mercaptans. The corrosive sulfur compounds have an obnoxious odor that is typically characterized as "rotten eggs".

Pyrophoric iron sulfide results from the corrosive action of sulfur compounds on the iron and steel used in refinery process equipment, piping, and tanks. The combustion of petroleum products containing sulfur compounds produces undesirables such as sulfuric acid and sulfur dioxide. Catalytic hydrotreating processes such as hydrodesulfurization remove sulfur compounds from refinery product streams. Sweetening processes either remove the obnoxious sulfur compounds or convert them to odorless disulfides, as in the case of mercaptans.

Oxygen compounds include phenols, ketones, and carboxylic acids. These occur in crude oils in varying amounts.

Nitrogen is found in lighter fractions of crude oil as basic compounds, and generally in heavier fractions of crude oil as nonbasic compounds that may also include trace metals. Trace metals found include copper, vanadium, and/or nickel. Nitrogen oxides can also form in process furnaces. The decomposition of nitrogen compounds in catalytic cracking and hydrocracking processes forms ammonia and cyanides that can cause corrosion.

Metals, including nickel, iron, and vanadium, are often found in crude oils in small quantities. Burning heavy fuel oils in refinery furnaces and boilers can leave deposits of vanadium oxide and nickel oxide in furnace boxes, ducts, and tubes. It is also desirable to remove trace amounts of arsenic, vanadium, and nickel prior to processing as they can poison industrial catalysts.

Crude oils also contain inorganic salts such as sodium chloride, magnesium chloride, and calcium chloride in suspension or dissolved in entrained water (brine). These salts must be removed or neutralized before processing to prevent catalyst poisoning, equipment corrosion, and fouling. Salt corrosion is caused by the hydrolysis of some metal chlorides to hydrogen chloride (HCl) and the subsequent formation of hydrochloric acid when crude is heated. Hydrogen chloride may also combine with ammonia to form ammonium chloride (NH_4Cl), which causes fouling and corrosion.

Carbon dioxide may result from the decomposition of bicarbonates present in or added to crude, or from steam used in the distillation process. Some crude oils contain naphthenic (organic) acids, which may become corrosive at temperatures above 450°F when the acid value of the crude is above a certain level.

Benzene and VOC emissions from wastes

The two main classes of petrochemical raw materials coming out of refineries are *olefins* (including ethylene and propylene) and *aromatics* (including benzene and xylene isomers), both of which are produced in very large quantities. Olefins are produced mainly in the steam cracking and catalytic reforming processes.

Aromatic hydrocarbons are mainly produced by catalytic reforming or similar processes. Benzene and one of its derivatives, xylene, are the main aromatics produced from the refining process. Benzene is mainly used as an additive in gasoline and as an intermediate to make other chemicals. It ranks in the top 20 chemicals for production volume. Smaller amounts of benzene are used to make various lubricants, explosives, and napalm. In addition to xylene, several benzene derivatives are in products we use every day. Examples are: styrene/ polystyrene (styrofoam and expanded polystyrene, which is used in consumer packaging and disposables, for home electronics and appliances; in adhesives, tires, cars, boats, asphalt, floor wax, kitchen countertops, film, CD jewel cases, toner, synthetic rubber); phenol (for pharmaceuticals, antiseptics, detergents, herbicides, in pesticides, dyes, pigments, in synthetic resin formulations); cylcohexane (for nylons and solvents).

The composition of benzene found in crude oil varies greatly even from the same region. It is nearly impossible to rigorously correlate benzene content by refinery size or sourcing; and in general the industry has not published such data for the general public. One Web reference source has shown that benzene levels in Alaskan crude can be as high as 7%.[2]

Currently the benzene content of gasoline is 1–2%. Because of its antiknock properties, a mixture of benzene-rich aromatics is added to gasoline as a replacement for alkyl lead compounds. Some of the benzene in the fuel is emitted from vehicles as unburned fuel.

Benzene is also formed as a partial combustion product of larger complex aromatic fuel components. Non-benzene aromatics in gasoline such as toluene, ethylbenzene, xylenes, and heavy reformate (C_9) tend to increase exhaust benzene levels.

Documents (*Baker et al.* v. *Chevron USA*) reviewed in a litigation against Chevron for an old Gulf refinery that was shut down in 1986 showed that there are numerous sources of benzene emissions from refinery waste streams. Examples of major sources of benzene emissions in refinery wastewater are desalter effluent, column overheads, sour water and tank drawdowns. Other sources of benzene emissions from wastewater sources include groundwater from historical spill incidents and quench water used in refineries that draw from contaminated groundwater. The relative contributions of the major sources vary by refinery due to the particular processing characteristics. The waste point of generation is generally determined when the waste enters the sewer at a refinery; a stream such as a stripped sour water can in one case be a waste, but in another case not considered a waste if it is hard-piped back to a desalter. Benzene concentrations found among the Chevron documents for different refinery waste streams are reported in Table 1.4.

The reported benzene concentrations in different waste streams in Table 1.4 were documented in the early 1990s. The authors have not been able to identify any open literature sources that report more recent measurements. The US EPA did perform its assessment at the same time period and applied a model to estimate benzene emissions, EPA reported that benzene fugitive emissions from plant sewers and downstream wastewater treatment works can range from 47 to 75% of the benzene found in refinery sewer systems. The same source document from which Table 1.4 was produced reported that two Exxon refineries averaged 27% benzene emissions based on material balances performed from the points of generation and the outlet of the API separator.

Refineries in the USA have generally gotten away without reporting many of their benzene and other fugitive emissions for a variety of reasons explained in Chapter 2. In waste streams the under-reporting of emissions has been a common practice since the early days when reporting requirements were first instituted. For example, slop oils from refineries are generally returned to the crude unit or coker for processing. Some refineries send slop oil back to the fluid

[2] http://www.alaskapub1-175lichealth.org/pdf/env/214.pdf

Table 1.4 Reported benzene concentrations in ppm in different refinery waste streams

Refinery	Desalter	Column overhead	Tank drawdowns
Alaska		80–93	3–12
El Paso	8–10	1–5	0–2
El Segundo	0–21	0–200	
Hawaii	1–175	600	1–200
Pascagoula	1–5	0–4	
Perth Amboy	900		1–900
Philadelphia	3–15	0–45	10–500
Port Arthur	16–18	200	5–100
Richmond	20–50	0–10	
Salt Lake	15–30	10–65	0–120

Source: Bates Stamp C1062023 0649931 – Carolyn Baker et al. v. Chevron, Civil Action No. 1:05CV227.

catalytic cracker (FCC). Slop oil is defined in the NESHAP regulations as the floating oil (and solids) that accumulate on the surface of an oil–water separator. It is subject to control if it contains more than 10 parts per million (ppm) benzene. But all refineries attempt to blend off-spec products whenever possible, including sometimes slop oil. In these situations refinery managers take the position that this practice does not constitute the handling of a waste, and hence fugitive emissions need not be reported. The industry in general has taken great steps to categorize many off-spec intermediate streams as non-waste categories in order to limit reporting.

Figures 1.11–1.13 are a series of matrices that the authors have compiled that summarize common emission sources from waste streams. Most of these streams are sources of fugitive emissions and in many, one would anticipate possible benzene and VOC emissions.

1.4.3   Products

The most important refinery product is motor gasoline, which is a blend of hydrocarbons with boiling ranges from ambient temperatures to about 400°F. The important qualities for gasoline are octane number (antiknock), volatility (starting and vapor lock), and vapor pressure (environmental control). Additives are used to enhance performance and provide protection against oxidation and rust formation.

Kerosene is a refined middle-distillate petroleum product that finds use as a jet fuel and in cooking and space heating. When used as a jet fuel, some of the critical qualities are freeze point, flash point, and smoke point. Commercial jet fuel has a boiling range of about 375–525°F, and military jet fuel 130–550°F. Kerosene, with less critical specifications, is used for lighting, heating, solvents, and blending into diesel fuel.

AIR EMISSIONS

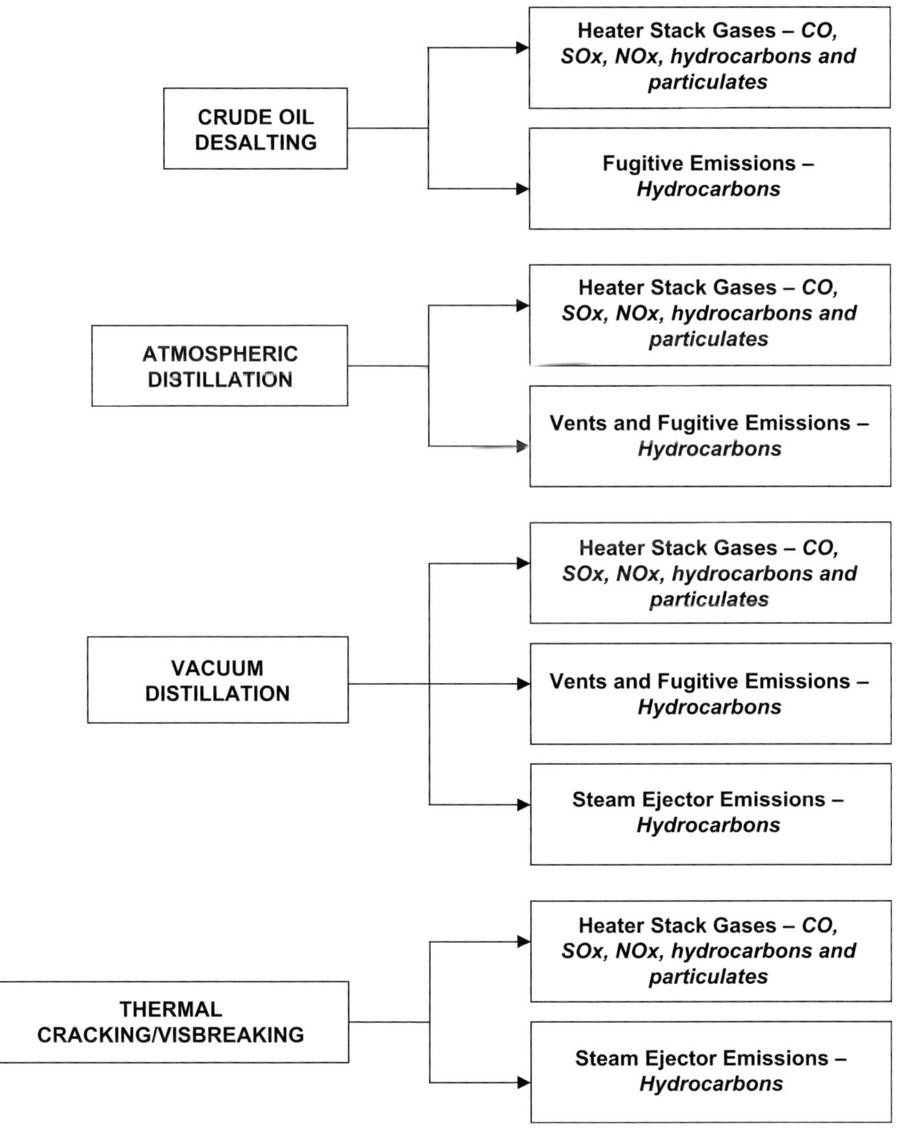

Figure 1.11 Air emissions summary.

AIR EMISSIONS

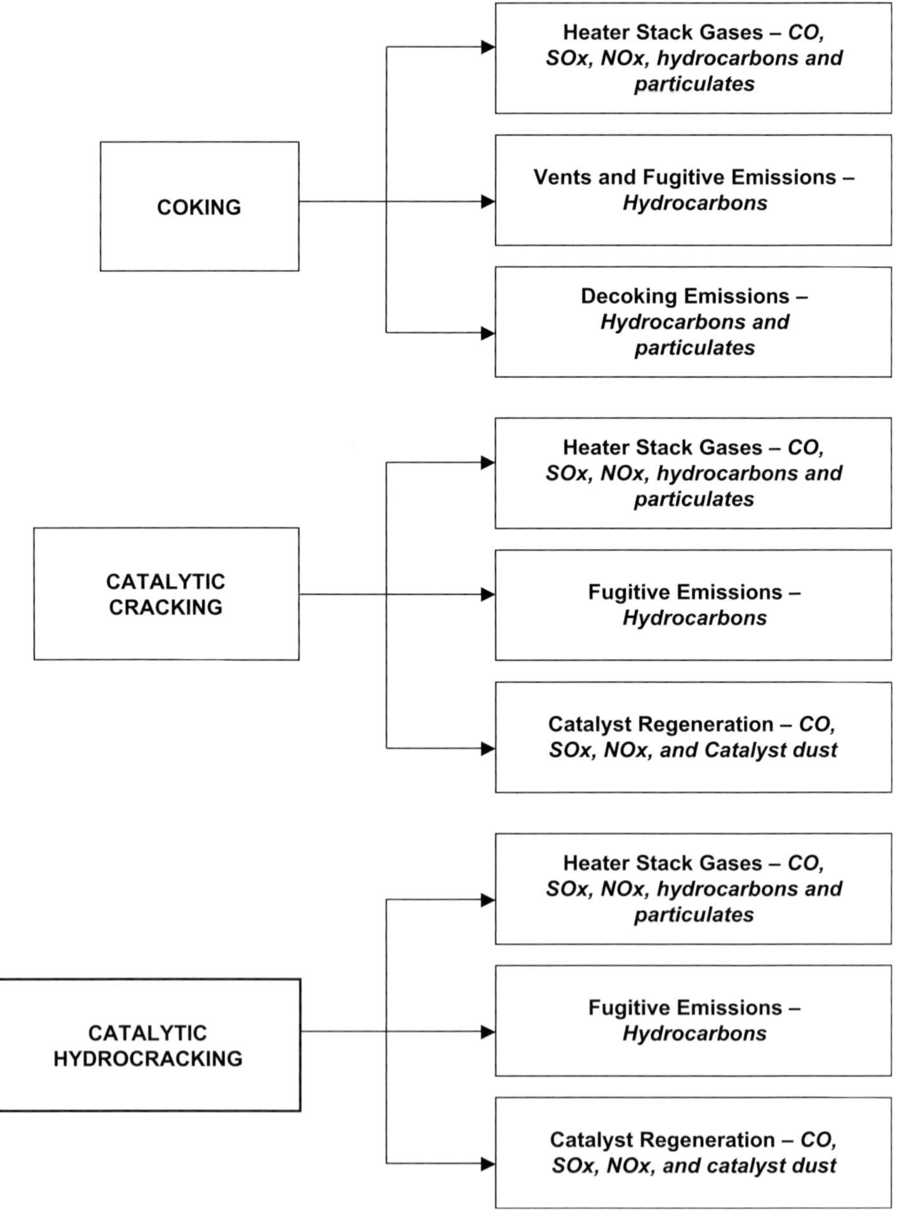

Figure 1.11 cont'd

AIR EMISSIONS

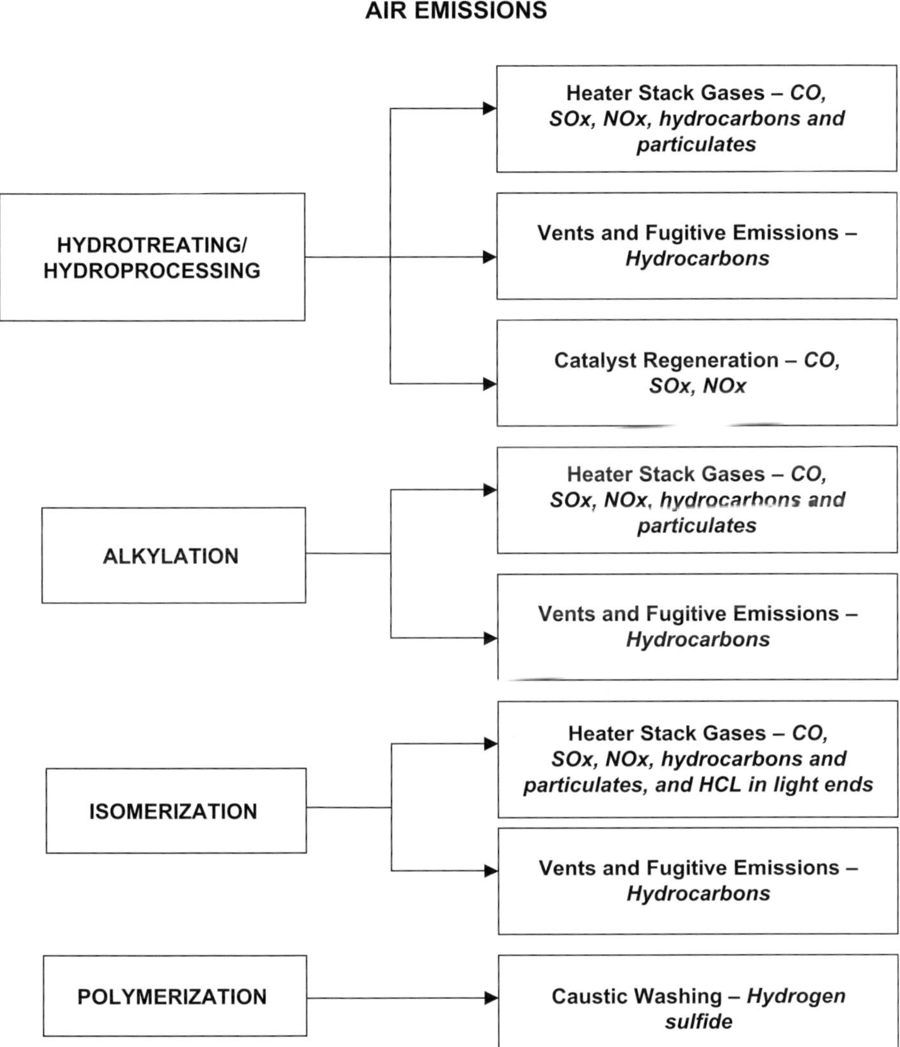

Figure 1.11 cont'd

Liquefied petroleum gas (LPG) consists principally of propane and butane and is produced for use as fuel and as an intermediate material in the manufacture of petrochemicals. The important specifications for proper performance include vapor pressure and control of contaminants.

Diesel fuels and domestic heating oils have boiling ranges of about 400–700°F. The desirable qualities required for distillate fuels include controlled flash and pour points, clean burning, no deposit formation in storage tanks, and a proper diesel fuel cetane rating for good starting and combustion.

AIR EMISSIONS

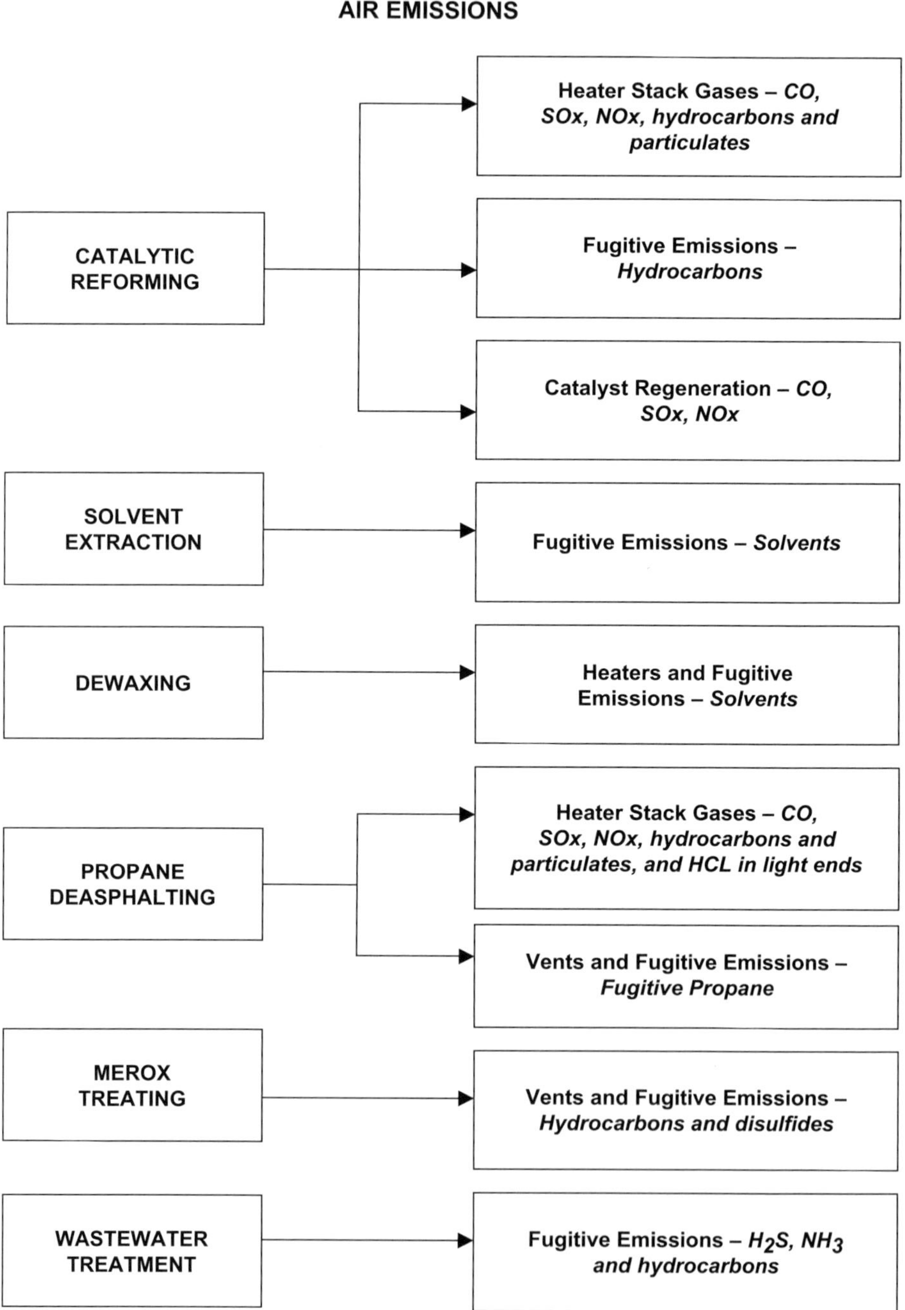

Figure 1.11 cont'd

PROCESS WASTEWATER

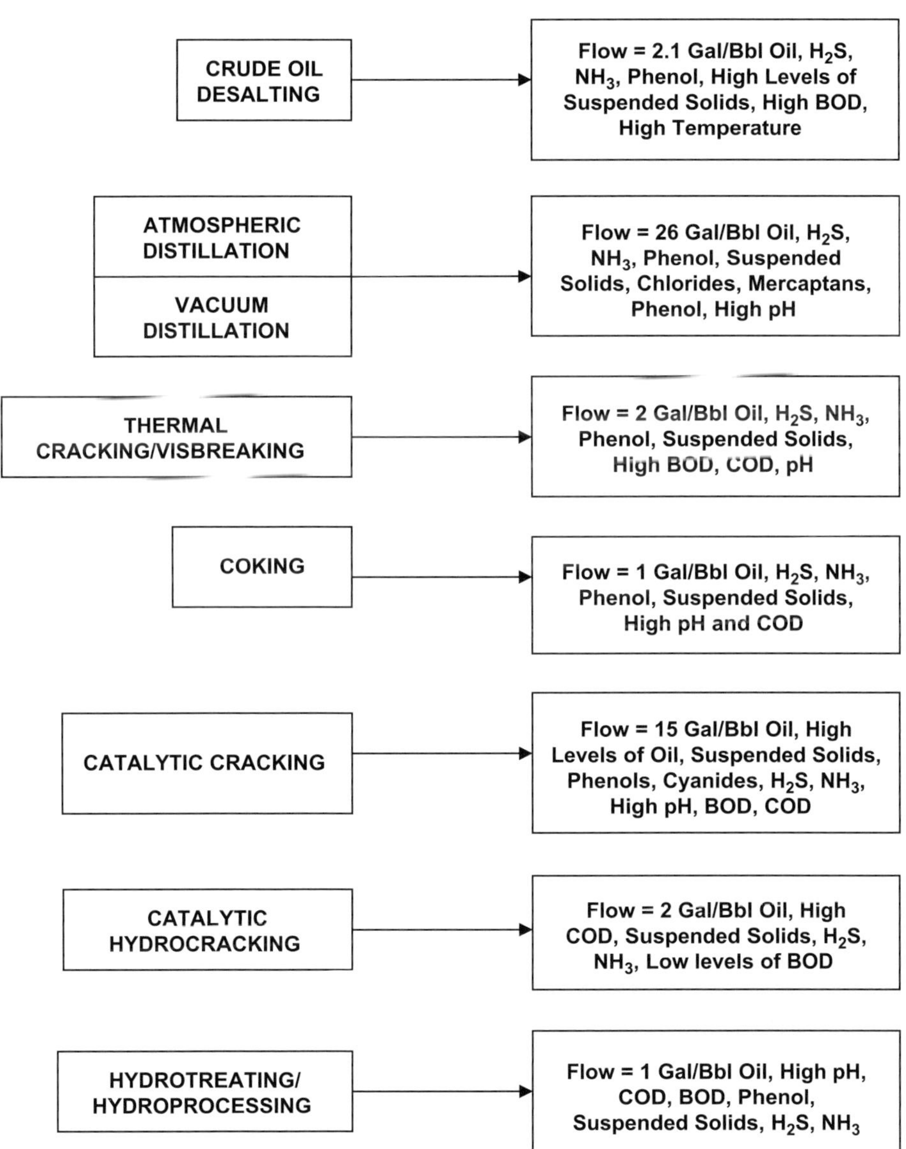

Figure 1.12 Process wastewater summary.

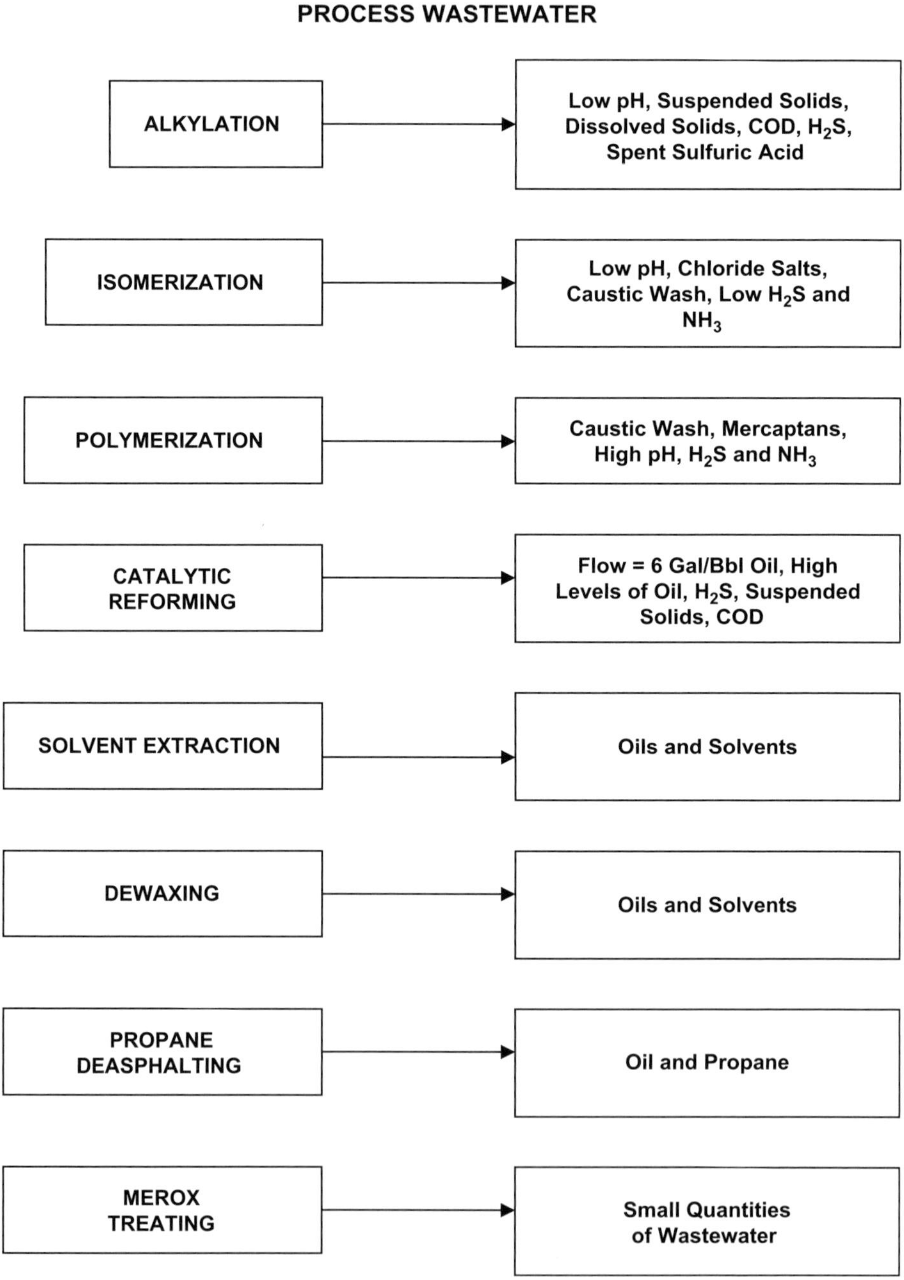

PROCESS WASTEWATER

| ALKYLATION | → | Low pH, Suspended Solids, Dissolved Solids, COD, H_2S, Spent Sulfuric Acid |

| ISOMERIZATION | → | Low pH, Chloride Salts, Caustic Wash, Low H_2S and NH_3 |

| POLYMERIZATION | → | Caustic Wash, Mercaptans, High pH, H_2S and NH_3 |

| CATALYTIC REFORMING | → | Flow = 6 Gal/Bbl Oil, High Levels of Oil, H_2S, Suspended Solids, COD |

| SOLVENT EXTRACTION | → | Oils and Solvents |

| DEWAXING | → | Oils and Solvents |

| PROPANE DEASPHALTING | → | Oil and Propane |

| MEROX TREATING | → | Small Quantities of Wastewater |

Figure 1.12 cont'd

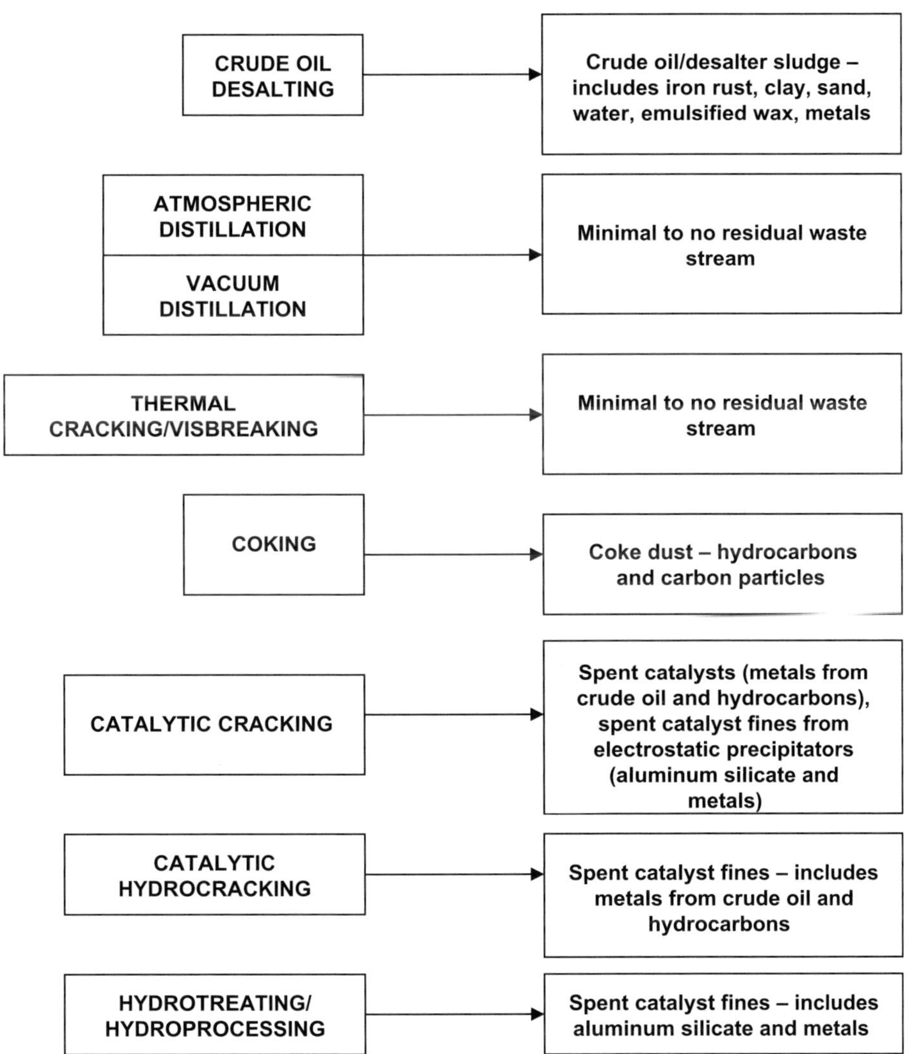

Figure 1.13 Residual waste streams summary.

Refineries also produce residual fuels for use by marine vessels, power plants, commercial buildings, and industry. These are combinations of residual and distillate fuels for heating and processing. The two most critical specifications of residual fuels are viscosity and low sulfur content for environmental control.

RESIDUAL WASTE STREAMS

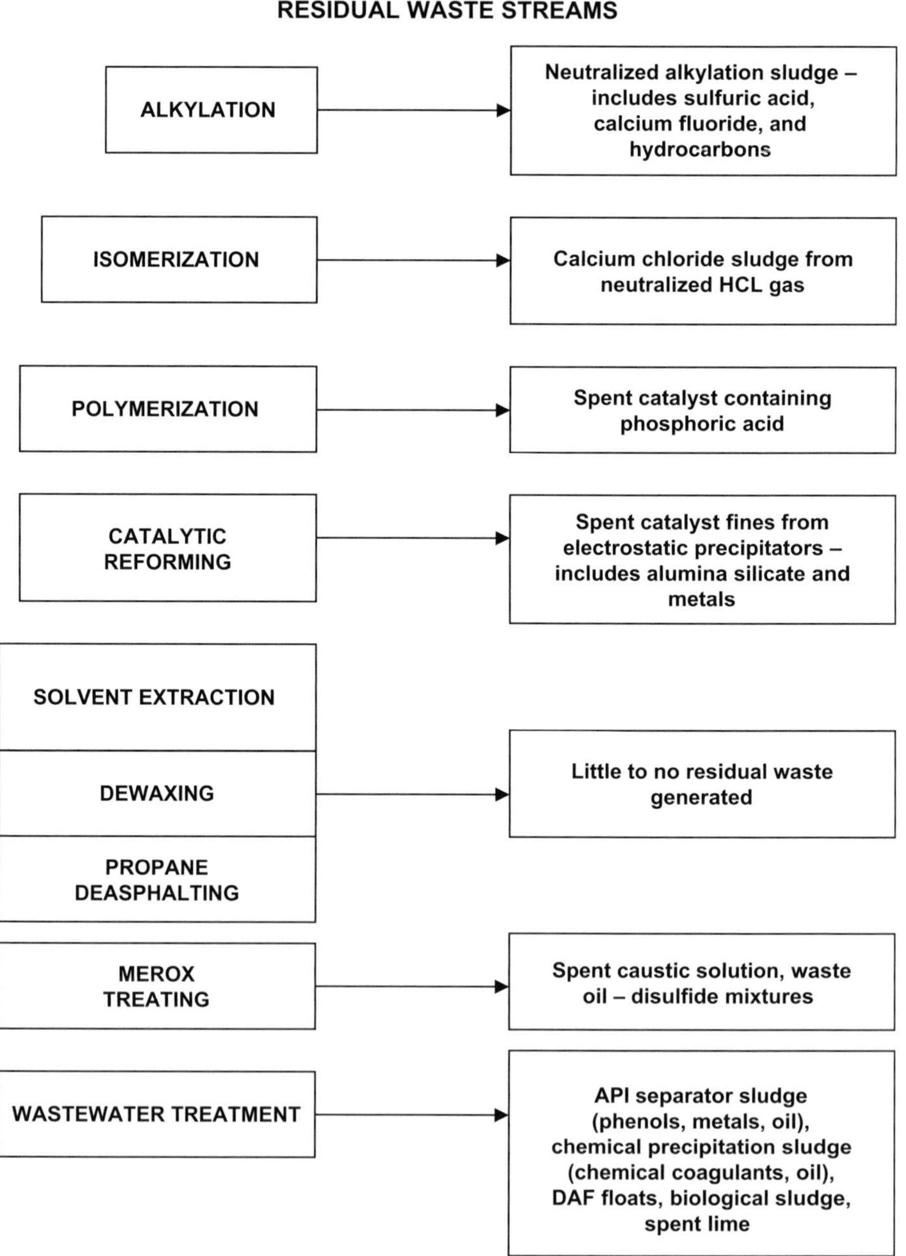

Figure 1.13 cont'd.

Refineries also make coke and asphalt. Coke is almost pure carbon with a variety of uses from electrodes to charcoal briquets. Asphalt, used for roads and roofing materials, must be inert to most chemicals and weather conditions.

Solvents are also made. A variety of products, whose boiling points and hydrocarbon composition are closely controlled, are produced for use as these materials. They include benzene, toluene, and xylene.

Many products derived from crude oil refining, such as ethylene, propylene, butylene, and isobutylene, are primarily intended for use as petrochemical feedstock in the production of plastics, synthetic fibers, synthetic rubbers, and other products.

Special refining processes produce lubricating-oil base stocks. Additives such as demulsifiers, antioxidants, and viscosity improvers are blended into the base stocks to provide the characteristics required for motor oils, industrial greases, lubricants, and cutting oils. The most critical quality for lubricating-oil base stock is a high viscosity index, which provides for greater consistency under varying temperatures.

A concern with fugitive emissions associated with the refined products is the presence of benzene, which is a confirmed human carcinogen. Benzene exists both in crude oil and gasoline; our comments are restricted to gasolines only.

Gasoline is a complex mixture of aliphatic and aromatic hydrocarbons derived from blending fractions of crude oil with brand-specific additives. The actual composition of any gasoline will vary according to the source of crude oil and the manufacturing process and between batches. Gasoline contains mixtures of volatile hydrocarbons and so inhalation is the most common form of exposure. Vapors can reach supralethal concentrations in confined or poorly ventilated areas (see reported studies by Aiden, 1958; Ainsworth, 1960; Takamiya et al., 2003). A representative sample of gasoline vapor concentrations under different exposure scenarios have been reported in the literature as follows:[3]

- 25,000 parts per million (ppm) in air above open barrel in unventilated outhouse on a "hot" day. Environmental conditions were not reported;
- 5–320 ppm in air around a tanker during bulk loading;
- 2–100 ppm in air around petrol pumps in a service station. Environmental conditions were not reported;
- 1–5 ppm in the air within a gasoline service station. The ambient temperature varied from 4.5 to 25°C. Recovered gasoline components were predominantly (72%) C_4 and C_5 aliphatic hydrocarbons;
- 174 ppm exposure for a worker at a bulk loading facility, 13 ppm exposure for a road tanker driver, and 4 ppm exposure for a service station worker. These are average exposure values reported. Environmental conditions were not reported. The vapor concentrations were reported as the sum of all detected hydrocarbon constituents.

[3] Petrol: Toxicological Review – see http://www.hpa.nhs.uk/web/HPAwebFile/HPAweb_C/ 1194947317038

As early as 1948 the American Petroleum Institute (API) published the guideline that the only absolutely safe level from exposure to benzene was zero. Despite this, in the USA, gasoline has contained an average of about 1.5% benzene for the past two decades, but has reportedly reached 5% by volume historically (Infante et al., 1990). Infante ct al. have noted that gasolines in most European countries have contained more benzene than US varieties over the years, with trends towards lower levels only occurring most recently.

Because gasoline station pumps do not provide adequate information on the cancers known to be associated with benzene exposure, and the material safety data sheets for gasoline do not provide the available evidence on chromosomal or genetic damage caused by benzene exposure, the public has not been adequately warned. The lack of candor about the hazards of benzene in gasoline has placed garage mechanics and highway maintenance workers at unnecessary risk by using gasoline as a solvent in cleaning auto parts. Consumers take unnecessary risks by using gasoline as a solvent and fail to take the necessary precautions when using gasoline in various home appliances such as lawn mowers, weed trimming devices, power saws, etc.

The hazards of benzene in gasoline have been recognized since at least the 1920s (Askey, 1928). Despite the overwhelming literature on the hazards of benzene in gasoline the industry has resisted providing adequate health and safety warnings to workers and consumers.

1.4.4 Refinery chemicals

There are a number of important chemicals that are used by refineries. These include leaded gasoline additives, oxygenates, caustics, sulfuric and hydrofluoric acids.

Tetra-alkyl lead

Tetraethyl lead (TEL) and tetramethyl lead (TML) are additives formerly used to improve gasoline octane ratings but are no longer in common use except in aviation gasoline. There are a few places throughout the world where leaded gasolines for public consumption are still manufactured; however, these are dwindling. The early history of the petroleum industry shows that the industry was not only acutely aware of the hazards associated with leaded gasoline, but in fact promoted a public relations campaign for decades to conceal health risks. See Kovarik (2005) for a historical perspective and the cover-up by the industry on the health risks of TEL.

Older readers are likely to recall the national debate over the safety of TEL in the 1960s. This debate came to a head when President Nixon signed the Clean Air Act into law. The Clean Air Act of 1970 mandated a 90% reduction in three major emissions: carbon monoxide, nitrogen oxides, and other hydrocarbons (mostly unburned fuel). This led to the introduction of the catalytic converter for automobiles, but in order for the catalytic converter to be used, lead had to be taken out of gasoline because it poisoned the platinum surfaces of the catalytic converter.

In 1973, the EPA announced regulations requiring a gradual reduction in the lead content of each refinery's total gasoline pool. At that time, the average gallon of gasoline contained 2.2 grams of lead. The lead phase-down started on 1 January 1975, with a reduction to 1.7 grams, and continued to 1979 with a reduction to 0.5 grams per gallon. Ethyl Corporation filed suit to keep lead in gasoline, and in late 1974, a panel of the US Court of Appeals for the District of Columbia Circuit set aside the EPA's lead regulations as "arbitrary and capricious", ruling in favor of Ethyl and DuPont. But the decision was reversed in 1976, when the full Court of Appeals cleared the way for a continued lead phase-down.

Automakers equipped new cars with pollution-reducing catalytic converters designed to run only on unleaded fuel beginning in 1975 and 1976, and new unleaded gasoline pumps began appearing at filling stations nationwide. By 1985, 40% of all gasoline sold was still leaded, but in July of that year, the refinery pool standard of 1.1 grams per gallon dropped to 0.5, then dropped further to 0.1 grams per gallon on 1 January 1986. The 1986 standard represented a drop of more than 98% in the lead content of US gasoline from 1970 to 1986.

Lead content in gasoline peaked in 1973 at an average of 2.2 grams per gallon, which amounted to about 200,000 tons of lead used per year in the USA. In 1995 leaded fuel accounted for only 0.6% of total gasoline sales and less than 2000 tons of lead per year. Effective from 1 January 1996, the Clean Air Act banned the sale of the small amount of leaded fuel that was still available in some parts of the country for use in on-road vehicles. Fuel containing lead was still permitted for some off-road uses, including aircraft, racing cars, farm equipment, and marine engines.

In 1996, the International Bank for Reconstruction and Development (the World Bank) recommended global phase-out of leaded gasoline. Following a phase-down period, in 2000 the European Economic Community also banned most leaded gasoline. Laws prohibiting leaded gasoline have been adopted worldwide in recent years.

Leaded gasoline is still being phased out in most developing nations. The "Declaration of Dakar", approved 28 June 2001, involved the World Bank and 25 sub-Saharan African nations in a plan to clean up the air quality in African cities. The most important part of the program was a phase-out of leaded gasoline.

Lead poisoning is one of the oldest known forms of occupational and environmental disease. Kovarik (2005) has noted that when scientists objected to the introduction of leaded gasoline in the 1920s, "they felt they had the obvious benefit of historical understanding. But deliberate miscalculations of the volume of leaded gasoline residues, political opposition, and positivistic attitudes about science meant that public health advocates could not block industry's use of lead in gasoline in the 1920s. Ethyl and the industries presented a very clear challenge to public health." These companies claimed that there were no alternatives to leaded gasoline, when in fact this was an outright lie.

Other oxygenates

Ethyl tertiary-butyl ether (ETBE), methyl tertiary-butyl ether (MTBE), tertiary-amyl methyl ether (TAME), and other oxygenates are employed to improve gasoline octane ratings and reduce carbon monoxide emissions. These additives have carried other environmental concerns, especially in groundwater contamination.

MTBE is a synthetic chemical compound that is manufactured through the chemical reaction of methanol and isobutylene (Agency for Toxic Substances and Diseases Registry, 2007). It is one of a group of chemicals commonly known as "oxygenates" because it raises the oxygen content of gasoline, which helps prevent the engine from "knocking". At room temperature, MTBE is a volatile, flammable, and colorless liquid that dissolves easily in water. It has been produced in very large quantities and is almost exclusively used as a fuel additive in unleaded gasoline in the USA and throughout the world. When it was introduced in the late 1970s, MTBE was added to premium-grade fuels in relatively low concentrations to increase the octane ratings. In the early 1990s, MTBE was added in much higher concentrations (up to 15%) to enhance fuel combustion and reduce tailpipe emissions (US Geological Survey, 2007). In 2007, over 21 million barrels of MTBE were produced in the USA (Energy Information Administration, 2008).

There has been significant controversy over the environmental threat from MTBE. MTBE may be released into the environment wherever gasoline is stored; however, leakage of underground fuel tanks and pipelines, and fuel spillage, account for the largest amount. There are opportunities for spillage whenever fuel is transported or transferred as well. Because MTBE has a low octanol–water partition coefficient (meaning it is attracted to water) compared to other organic gasoline components (e.g. benzene, toluene, ethylbenzene, and xylene – known collectively as BTEX compounds), MTBE will travel rapidly through soil into groundwater (Hinck, 2001). Gasoline containing MTBE that leaks, spills, or is otherwise released into the environment will quickly reach the water table and will contaminate wells that draw from the affected underground aquifers, whereas BTEX compounds tend to bind to soil and migrate slowly to the water.

MTBE has been detected in the air and water. It quickly evaporates from open containers and surface water, so it is commonly found as vapor in air. The ATSDR (2007) has reported that small amounts of MTBE may dissolve in water when MTBE levels in the water are below those found in the air, and this may travel to underground aquifers. Groundwater used for drinking in the Santa

Monica area has been reported to have up to 610 parts per billion (ppb) MTBE concentrations (US EPA, 2008).

MTBE is not readily biodegradable underground, and hence MTBE that is released into groundwater has been shown to linger for many decades. As such, in 1999, the US Geological Survey detected MTBE in 21% of groundwater samples taken from areas chosen specifically because there were no records that gasoline had ever leaked or been spilled there (Hinck, 2001).

Caustics

Caustics are added to desalting water to neutralize acids and reduce corrosion. They are also added to desalted crude in order to reduce the amount of corrosive chlorides in the tower overheads. They are used in some refinery treating processes to remove contaminants from hydrocarbon streams.

Sulfuric acid and hydrofluoric acid are used primarily as catalysts in alkylation processes. Sulfuric acid is also used in some treatment processes.

1.5   Refining technologies

1.5.1   Classification of technologies

Figure 1.14 shows a generalized process flow chart for a typical refinery. Table 1.5 provides a chronological summary of the refining technologies and years in which they were introduced commercially. Major sections of refineris include three major groups:

- Fractionation
- Conversion
  - Decomposition
  - Unification
  - Rearrangement
- Treatment.

These are briefly discussed.

Distillation processes

The first US refinery was commissioned in 1861. It produced kerosene by means of atmospheric distillation. By-products included tar and naphtha. It was soon discovered that high-quality lubricating oils could be produced by distilling petroleum under a vacuum. However, for the next 30 years kerosene was the product of demand in the market. The invention of the electric light decreased the demand for kerosene, but the invention of the internal combustion engine created a demand for diesel fuel and gasoline (naphtha).

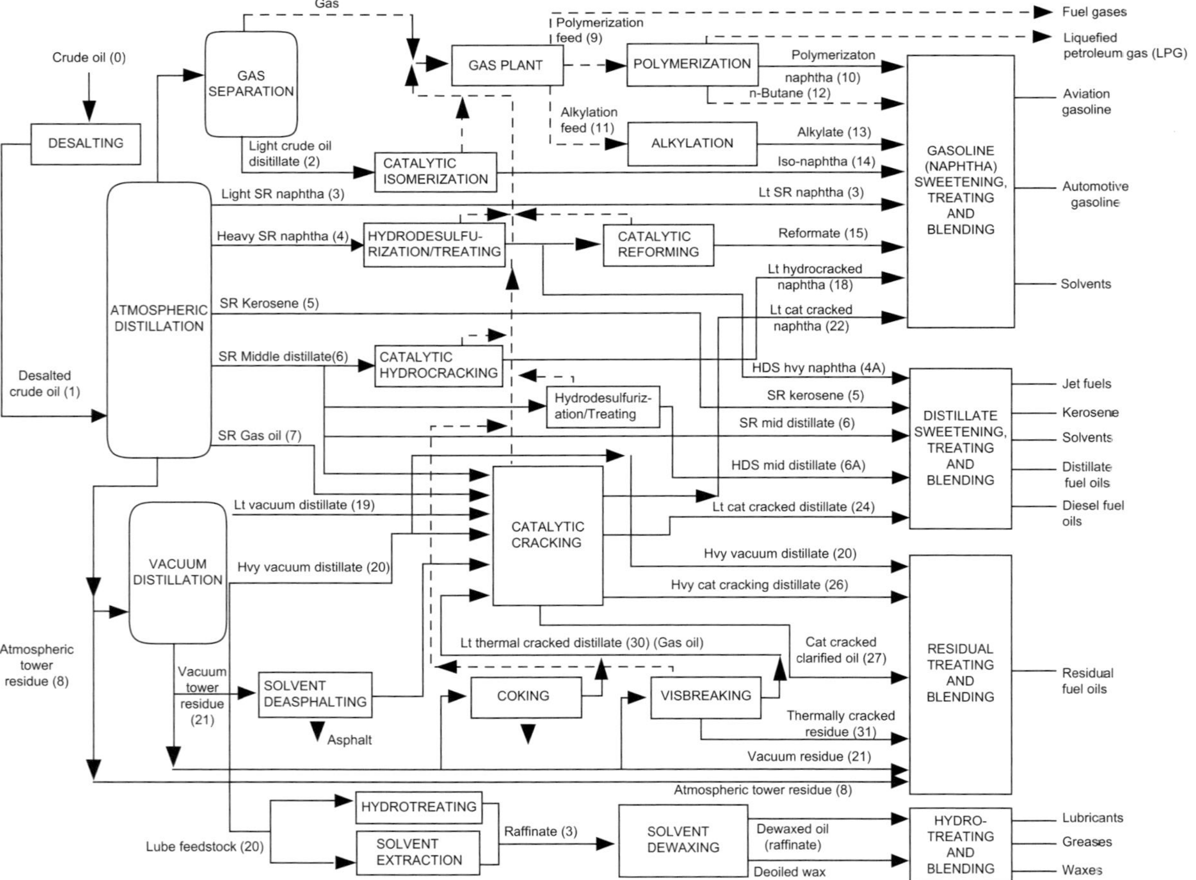

Figure 1.14 Generalized flow chart of a typical refinery.
Source: http://www.osha.gov/dts/osta/otm/otm_iv/otm_iv_2.html

Table 1.5 Commercialization of major refining technologies

Year	Process name	Purpose	By-products, etc.
1862	Atmospheric distillation	Produce kerosene	Naphtha, tar, etc.
1870	Vacuum distillation	Lubricants (original) Cracking feedstocks (1930s)	Asphalt, residual coker feedstocks
1913	Thermal cracking	Increase gasoline	Residual, bunker fuel
1916	Sweetening	Reduce sulfur and odor	Sulfur
1930	Thermal reforming	Improve octane number	Residual
1932	Hydrogenation	Remove sulfur	Sulfur
1932	Coking	Produce gasoline base stocks	Coke
1933	Solvent extraction	Improve lubricant viscosity index	Aromatics
1935	Solvent dewaxing	Improve pour point	Waxes
1935	Cat. polymerization	Improve gasoline yield and octane number	Petrochemical feedstocks
1937	Catalytic cracking	Higher octane gasoline	Petrochemical feedstocks
1939	Visbreaking	Reduce viscosity	Increased distillate, tar
1940	Alkylation	Increase gasoline octane and yield	High-octane aviation gasoline
1940	Isomerization	Produce alkylation feedstock	Naphtha
1942	Fluid catalytic cracking	Increase gasoline yield and octane	Petrochemical feedstocks
1950	Deasphalting	Increase cracking feedstock	Asphalt
1952	Catalytic reforming	Convert low-quality naphtha	Aromatics
1954	Hydrodesulfurization	Remove sulfur	Sulfur
1956	Inhibitor sweetening	Remove mercaptan	Disulfides
1957	Catalytic isomerization	Convert to molecules with high octane number	Alkylation feedstocks
1960	Hydrocracking	Improve quality and reduce sulfur	Alkylation feedstocks
1974	Catalytic dewaxing	Improve pour point	Wax
1975	Residual hydrocracking	Increase gasoline yield from residual	Heavy residuals

Thermal cracking processes

World War I created the impetus for mass production whereby the number of gasoline-powered vehicles increased dramatically. Subsequently, the demand for gasoline grew accordingly. However, distillation processes produced only a certain amount of gasoline from crude oil.

In 1913, the thermal cracking process was developed, which subjected heavy fuels to both pressure and intense heat, physically breaking the large molecules into smaller ones to produce additional gasoline and distillate fuels. Visbreaking

Table 1.6 OSHA quick guide to refinery processes

Process name	Action	Method	Purpose	Feedstock(s)	Product(s)
Fractionation processes					
Atmospheric distillation	Separation	Thermal	Separate fractions	Desalted crude oil	Gas, gas oil, distillate, residual
Vacuum distillation	Separation	Thermal	Separate w/o cracking	Atmospheric tower residual	Gas oil, lube stock, residual
Conversion processed – decomposition					
Catalytic cracking	Alteration	Catalytic	Upgrade gasoline	Gas oil, coke distillate	Gasoline, petrochemical feedstock
Coking	Polymerize	Thermal	Convert vacuum residuals	Gas oil, coke distillate	Gasoline, petrochemical feedstock
Hydrocracking	Hydrogenate	Catalytic	Convert to lighter HCs	Gas oil, cracked oil, residual	Lighter, higher-quality products
*Hydrogen steam reforming	Decompose	Thermal/catalytic	Produce hydrogen	Desulfurized gas, O_2, steam	Hydrogen, CO, CO_2
*Steam cracking	Decompose	Thermal	Crack large molecules	Atm. tower heavy fuel/distillate	Cracked naphtha, coke, residual
Visbreaking	Decompose	Thermal	Reduce viscosity	Atmospheric tower residual	Distillate, tar
Conversion processes – unification					
Alkylation	Combining	Catalytic	Unite olefins and isoparaffins	Tower isobutane/ cracker olefin	Iso-octane (alkylate)
Grease compounding	Combining	Thermal	Combine soaps and oils	Lube oil, fatty acid, alky metal	Lubricating grease
Polymerizing	Polymerize	Catalytic	Unite 2 or more olefins	Cracker olefins	High-octane naphtha, petrochemical stocks

Conversion processes – alteration or rearrangement

Catalytic reforming	Alteration/ dehydration	Catalytic	Upgrade low-octane naphtha	Coker/hydrocracker naphtha	High-octane reformate/aromatic
Isomerization	Rearrange	Catalytic	Convert straight chain to branch	Butane, pentane, hexane	Isobutane/pentane/ hexane

Treatment processes

*Amine treating	Treatment	Absorption	Remove acidic contaminants	Sour gas, HCs w/CO_2 and H_2S	Acid-free gases and liquid HCs
Desalting	Dehydration	Absorption	Remove contaminants	Crude oil	Desalted crude oil
Drying and sweetening	Treatment	Abspt./therm.	Remove H_2O and sulfur compounds	Liq. HCs, LPG, alky. feedstock	Sweet and dry hydrocarbons
*Furfural extraction	Solvent extr.	Absorption	Upgrade middle distillate and lubes	Cycle oils and lube feedstocks	High-quality diesel and lube oil
Hydrodesulfurization	Treatment	Catalytic	Remove sulfur, contaminants	High-sulfur residual/ gas oil	Desulfurized olefins
Hydrotreating	Hydrogenation	Catalytic	Remove impurities, saturate HCs	Residuals, cracked HCs	Cracker feed, distillate, lube
*Phenol extraction	Solvent extr.	Abspt./therm.	Improve visc. index, color	Lube oil base stocks	High-quality lube oils
Solvent deasphalting	Treatment	Absorption	Remove asphalt	Vac. tower residual, propane	Heavy lube oil, asphalt
Solvent dewaxing	Treatment	Cool/filter	Remove wax from lube stocks	Vac. tower lube oils	Dewaxed lube basestock
Solvent extraction	Solvent extr.	Abspt./precip.	Separate unsat. oils	Gas oil, reformate, distillate	High-octane gasoline
Sweetening	Treatment	Catalytic	Remove H_2S, convert mercaptan	Untreated distillate/ gasoline	High-quality distillate/gasoline

was a major technology innovation in the 1930s. It is a form of thermal cracking capable of producing more desirable and valuable products.

Catalytic processes

Higher-compression gasoline engines required higher-octane gasoline with better antiknock characteristics. The introduction of catalytic cracking and polymerization processes in the mid to late 1930s met the demand by providing improved gasoline yields and higher octane numbers. Alkylation is another catalytic process that was developed in the early 1940s, to produce more high-octane aviation gasoline and petrochemical feedstock for explosives and synthetic rubber.

Catalytic isomerization was developed to convert hydrocarbons to produce increased quantities of alkylation feedstock. Improved catalysts and process methods such as hydrocracking and reforming were developed throughout the 1960s to increase gasoline yields and improve antiknock characteristics. These catalytic processes also produced hydrocarbon molecules with a double bond (alkenes) and formed the basis of the modern petrochemical industry.

Treatment processes

Various treatment methods have always been used to remove non-hydrocarbons, impurities, and other constituents that adversely affect the properties of finished products or reduce the efficiency of the conversion processes. Treating involves chemical reaction and/or physical separation. Typical examples of treating are chemical sweetening, acid treating, clay contacting, caustic washing, hydrotreating, drying, solvent extraction, and solvent dewaxing. Sweetening compounds and acids desulfurize crude oil before processing and treat products during and after processing. Following the Second World War, various reforming processes were developed to improve gasoline quality and yield and to produce higher-quality products. Some of these involved the use of catalysts and/or hydrogen to change molecules and remove sulfur. A number of the more commonly used treating and reforming processes are described later.

Formulating and blending

Formulating and blending is the process of mixing and combining hydrocarbon fractions, additives, and other components to produce finished products with specific performance properties.

Other refining operations

Other important refining operations include: light-ends recovery; sour-water stripping; solid waste and wastewater treatment; process-water treatment and cooling; storage and handling; product movement; hydrogen production; acid and tail-gas treatment; and sulfur recovery.

Auxiliary operations and facilities include: steam and power generation; process and fire water systems; flares and relief systems; furnaces and heaters; pumps and valves; supply of steam, air, nitrogen, and other plant gases; alarms and sensors; noise and pollution controls; sampling, testing, and inspecting; and laboratory, control room, maintenance, and administrative facilities.

1.5.2 Pretreatment

One form of crude oil pretreatment is known as "desalting". Crude oil usually contains water, inorganic salts, suspended solids, and water-soluble trace metals. As a first step in the refining process, to reduce corrosion, plugging, and fouling of equipment and to prevent poisoning the catalysts in processing units, these contaminants are removed by the process of dehydration, which is referred to as desalting.

Two methods of crude-oil desalting are chemical and electrostatic separation. Both use hot water as the extraction agent. In chemical desalting, water and chemical surfactant (demulsifiers) are added to the crude, followed by heating so that salts and other impurities dissolve into the water; the resultant mixture is then held in a tank where the impurities settle out.

Electrostatic desalting involves the application of high-voltage electrostatic charges to concentrate suspended water globules in the bottom of the settling tank. Surfactants are added only when the crude has a large amount of suspended solids. Both methods of desalting are continuous. Figure 1.15 shows the essential components in desalting.

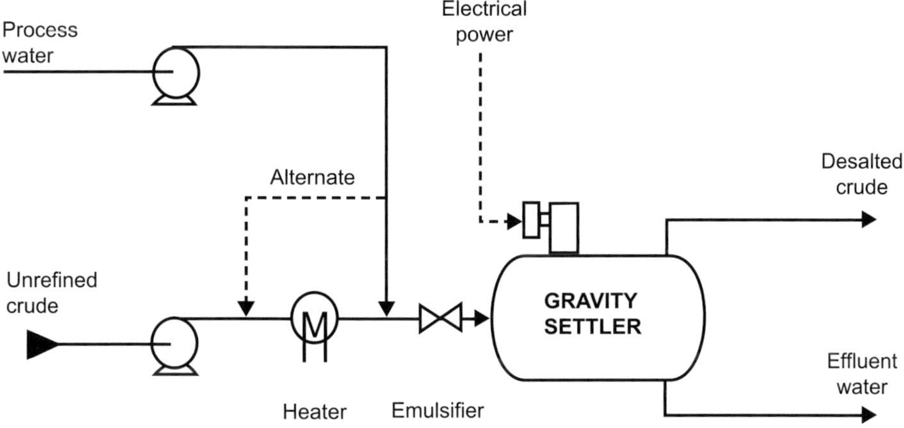

Figure 1.15 Simple schematic of electrostatic desalting.

A third and less-common process involves filtering heated crude using diatomaceous earth.

The feedstock crude oil is heated to between 150 and 350°F to reduce viscosity and surface tension for easier mixing and separation of the water. The temperature is limited by the vapor pressure of the crude-oil feedstock. In both methods other chemicals may be added. Ammonia is often used to reduce corrosion. Caustic or acid may be added to adjust the pH of the water wash. Wastewater and contaminants are discharged from the bottom of the settling tank to the wastewater treatment facility. The desalted crude is continuously drawn from the top of the settling tanks and sent to the crude distillation (fractionating) tower.

The potential exists for a fire due to a leak or release of crude from heaters in crude desalting units. Low-boiling-point components of crude may also be released if a leak occurs. Fugitive emissions can result from loose fittings, leaking valves, couplings, and joints.

Inadequate desalting can result in the fouling of heater tubes and heat exchangers throughout the refinery. Fouling restricts product flow and heat transfer and leads to failures due to increased pressures and temperatures. Corrosion, which occurs due to the presence of hydrogen sulfide, hydrogen chloride, naphthenic (organic) acids, and other contaminants in the crude oil, also causes equipment failure, which in turn can result in spills, leaks, and fugitive emissions. Neutralized salts (ammonium chlorides and sulfides), when moistened by condensed water, can cause corrosion. Overpressuring the unit is another potential hazard that causes failures and can lead to catastrophic releases.

Desalting is designed as a closed process, and hence the greatest concern from an environmental standpoint is a leak or unplanned releases. Where elevated operating temperatures are used when desalting sour crudes, hydrogen sulfide will be present. There is the possibility of releases of ammonia, dry chemical demulsifiers, caustics, and/or acids.

From a worker safety standpoint, workers must adhere to strict safe work practices along with the use of appropriate personal protective equipment in order to minimize exposures to chemicals and other hazards such as heat. Workers are required to conduct process sampling, inspection, maintenance, and various turnaround activities for maintenance and cleaning purposes.

Depending on the crude feedstock and the treatment chemicals used, process wastewater will contain chlorides, sulfides, bicarbonates, ammonia, hydrocarbons, phenol, and suspended solids. If diatomaceous earth is used in the filtration process, exposures should be minimized or controlled since diatomaceous earth can contain silica in very fine particle size, making this a potential respiratory hazard.

1.5.3 Crude-oil distillation

Crude-oil distillation is also known as *fractionation*. The first step in the refining process is the separation of crude oil into various fractions or straight-run cuts by distillation in an atmospheric distillation column. The main fractions or "cuts" obtained have specific boiling-point ranges and can be classified in order of decreasing volatility into gases, light distillates, middle distillates, gas oils, and residuum.

The following is a brief description of an *atmospheric distillation tower*. The desalted crude feedstock is first preheated using recovered process heat. The feedstock is then sent to a direct-fired crude charge heater, where it is fed into the vertical distillation column just above the bottom. Here the pressures are slightly above atmospheric and temperatures range from 650 to 700°F. Any heating of crude oil above these temperatures can cause undesirable thermal cracking.

All but the heaviest fractions flash into vapor. As the hot vapor rises in the tower, temperatures are reduced via cooling by natural convection. Heavy fuel oil or asphalt residue is taken from the bottom. At successively higher points on the tower, the various major products including lubricating oil, heating oil, kerosene, gasoline, and uncondensed gases (which condense at lower temperatures) are captured.

The fractionating tower is essentially a steel cylinder that is typically about 120 feet tall. Inside the shell are horizontal steel trays that are used for separating and collecting the various liquid cuts. At each tray, vapors from below enter through perforations and bubble caps. They permit the vapors to bubble through the liquid on the tray, causing some condensation at the temperature of that tray. An overflow pipe drains the condensed liquids from each tray back to the tray below, where the higher temperature causes re-evaporation. Figure 1.16 shows a photograph of a refinery in the Middle East where distillation columns can be seen.

Figure 1.17 shows a simplified process flow sheet of crude oil distillation.

The evaporation, condensing, and scrubbing operation is repeated many times until the desired degree of product purity is achieved. Side streams from certain trays are taken off to obtain the desired fractions. Products ranging from uncondensed fixed gases at the top to heavy fuel oils at the bottom can be taken continuously from a fractionating tower. Steam is used in towers to lower the vapor pressure and create a partial vacuum. The distillation process separates the major constituents of crude oil into what is referred to as straight-run products. Sometimes crude oil is "topped" by distilling off only the lighter fractions, leaving a heavy residue that is often distilled further under high vacuum.

Recapping, the crude first goes through the desalting process. The desalted crude then goes to separation (fractionation) for straight-run products in an atmospheric distillation column. Typical products include naphthas, kerosene or distillates, gas oil, and residual. Naphthas are sent on for reforming (treating) as well as the kerosene and distillates. Gas will be sent to the catalytic cracking unit described later. Residuals are sent to a vacuum distillation tower or visbreaker.

A *vacuum distillation tower* is employed to further distill the residuum or topped crude from the atmospheric tower at higher temperatures. Reduced

Figure 1.16 Distillation columns can be seen in the left and right portions of the photo.

pressure is required to prevent thermal cracking from occurring. Normally the process takes place in one or more vacuum distillation towers. The principles of vacuum distillation operation resemble those of fractional distillation. A difference lies in that a vacuum distillation unit has a larger-diameter column that serves to maintain comparable vapor velocities at the reduced pressures. The internal designs of vacuum towers can vary from simple tray configurations like atmospheric columns to the use of random packing and demister pads instead of trays. A first-phase vacuum tower produces gas oils, lubricating-oil base stocks, and heavy residual for propane deasphalting. A second-phase tower operating at lower vacuum is used to distill surplus residuum from the atmospheric tower, which is not used for lube-stock processing, and surplus residuum from the first vacuum tower not used for deasphalting. Vacuum towers are used to separate catalytic cracking feedstock from surplus residuum. Figure 1.18 shows a simplified process flow sheet for vacuum distillation.

In addition to atmospheric and vacuum distillation columns, refineries rely on other distillation towers, which are simply referred to as *columns*. These are smaller distillation towers that are designed to separate specific products. Columns all work on the same principles as the towers described above. For example, a depropanizer is a small column designed to separate propane and lighter gases from butane and heavier components. Another larger column is

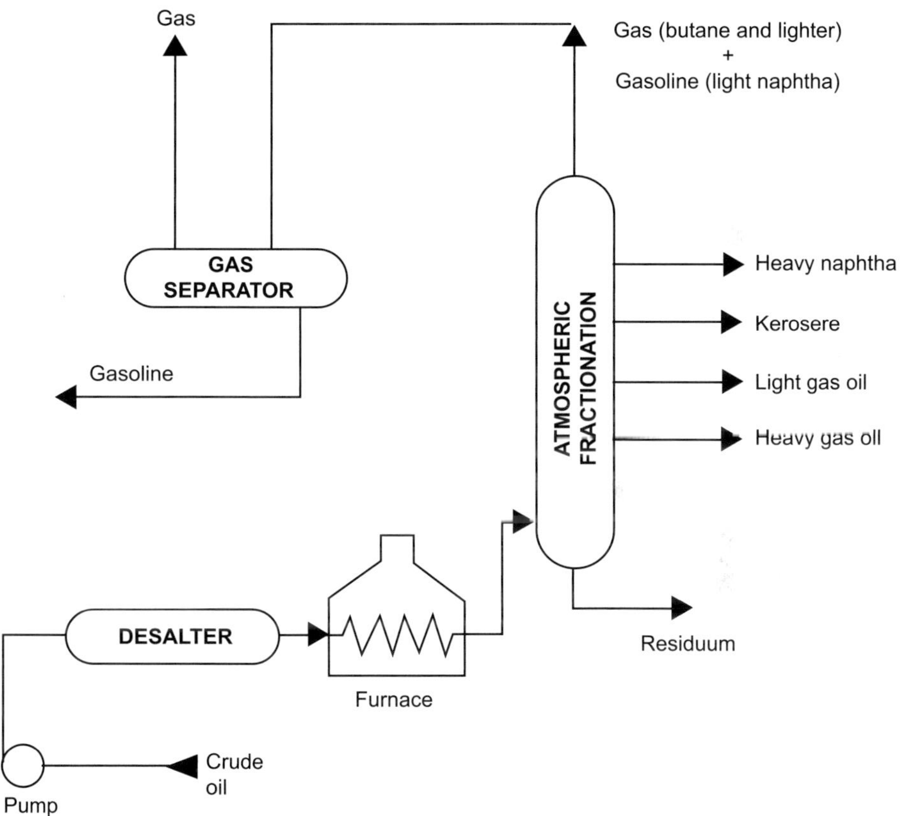

Figure 1.17 Simple schematic of atmospheric distillation.

used to separate ethylbenzene and xylene. Small "bubble" towers called strippers use steam to remove trace amounts of light products from the heavier product streams.

All of the above processes are designed as closed systems; however, there are numerous heaters and exchangers in the atmospheric and vacuum distillation units that are sources of ignition, spills, leaks, and fugitive emissions. Excursions in pressure, temperature, or liquid levels may occur if automatic controls fail. Control of temperature, pressure, and reflux within operating parameters is needed to prevent thermal cracking within the towers. Relief systems must be used for overpressure and operations monitored to prevent the crude from entering the reformer charge.

Sections most susceptible to corrosion from HCl and H_2S include the preheat exchanger, preheat furnace and bottoms exchanger from H_2S and sulfur compounds, atmospheric tower and vacuum furnace from H_2S, sulfur compounds, and organic acids, vacuum tower from H_2S and organic acids,

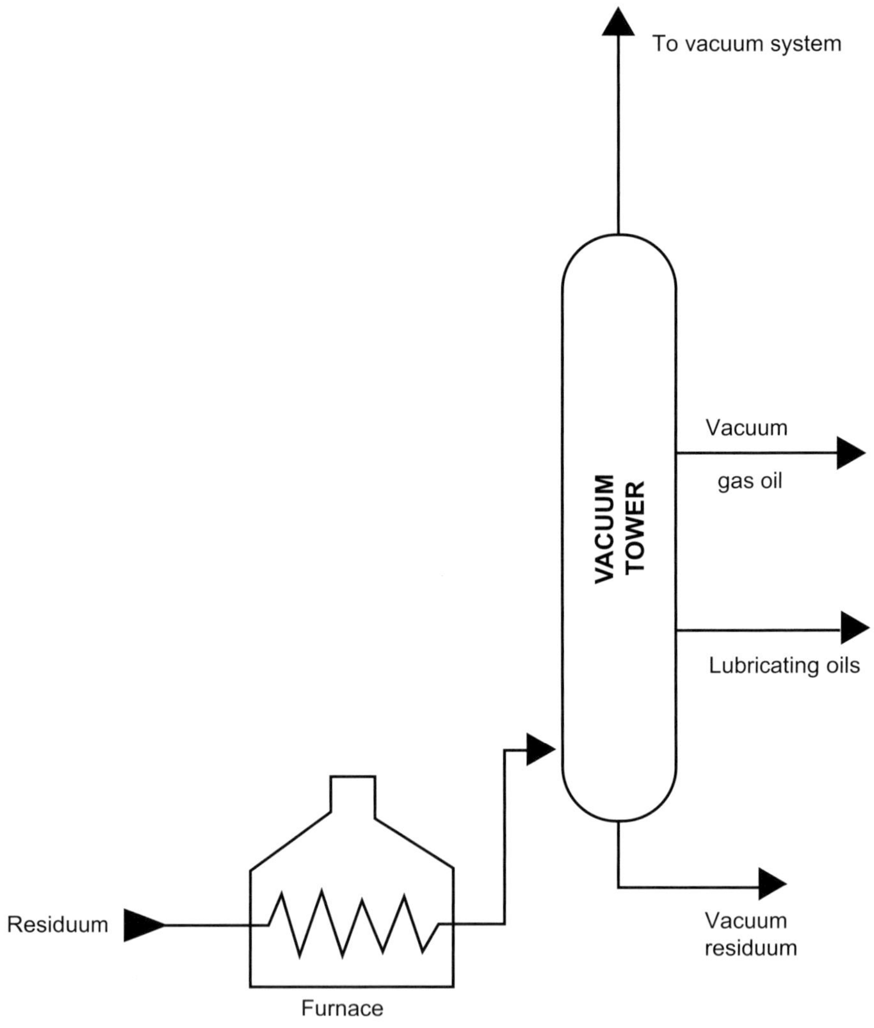

Figure 1.18 Simple schematic of the vacuum distillation process.

and overhead from H_2S, HCl, and water. Where sour crudes are processed, severe corrosion can occur in furnace tubing and in both atmospheric and vacuum towers where metal temperatures exceed 450°F. Wet H_2S also will cause cracks in steel leading to spills, leaks, and fugitive emissions. When processing high-nitrogen crudes, nitrogen oxides can form in the flue gases of furnaces. Nitrogen oxides are corrosive to steel when cooled to low temperatures in the presence of water. Aging refineries that have been in service for decades have fugitive emissions from these sources that the authors do not believe are adequately accounted for. Chemicals are used to control corrosion by HCl produced in distillation units. Ammonia may be injected into the overhead stream prior to initial condensation and/or an alkaline solution may be injected into the hot crude-oil feed. If sufficient wash water is not injected, deposits of ammonium chloride can form causing serious corrosion. Crude feedstock may contain appreciable amounts of water in suspension. This water phase can separate during startup and, along with water remaining in the tower from steam purging, settle in the bottom of the tower. When the water is heated to its boiling point an instantaneous vaporization explosion can occur upon contact with the oil in the unit. This can lead to devastating damage to the columns and catastrophic release of vapors and liquid.

1.5.4 Solvent extraction

Solvent treating is a process used for refining lubricating oils and other refinery stocks. Since distillation separates petroleum products into groups only by their boiling-point ranges, impurities remain. The impurities consist of organic compounds containing sulfur, nitrogen, and oxygen, inorganic salts and dissolved metals, and soluble salts that are present in the crude feedstock. In addition, kerosene and distillates usually have trace amounts of aromatics and naphthenes, and lubricating-oil base stocks contain wax.

Solvent refining processes including solvent extraction and solvent dewaxing. These methods remove the impurities at intermediate refining stages or just before sending the product to storage.

The role of solvent extraction is to prevent corrosion, protect catalyst in subsequent processes, and improve finished products by removing unsaturated, aromatic hydrocarbons from lubricant and grease stocks. The process separates aromatics, naphthenes, and impurities from the product stream by dissolving or precipitation. The feedstock is first dried and then treated using a continuous countercurrent solvent treatment operation. There are several variations of the method. In one variant the feedstock is washed with a liquid in which the substances to be removed are more soluble than in the desired resultant product. In another process, selected solvents are added to cause impurities to precipitate out of the product. In the adsorption process, highly porous solid materials collect liquid molecules on their surfaces. Solvent is separated from the product stream by heating, evaporation, or fractionation. The residual trace amounts are subsequently removed from the raffinate by steam stripping or vacuum flashing. Electrostatic precipitation may be used for separation of inorganic compounds. The solvent is then regenerated to be used again in the process. Common extraction solvents are phenol, furfural, and cresylic acid. Other solvents used are liquid sulfur dioxide, nitrobenzene, and 2,2′-dichloroethyl ether. The selection of specific processes and chemical agents depends on the nature of the feedstock being treated, the contaminants present, and the finished product requirements. A simplified process flow sheet for aromatics extraction is shown in Figure 1.19.

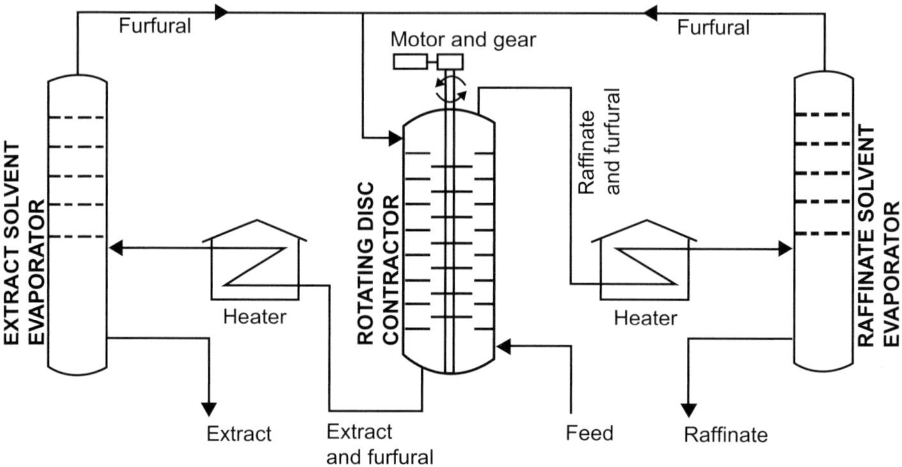

Figure 1.19 Simple schematic of aromatics extraction.

1.5.5 Solvent dewaxing

Solvent dewaxing is used to remove wax from distillate or residual base stocks at any stage in the refining process. While there are several solvent dewaxing processes all have the same general steps:

1. mixing the feedstock with a solvent;
2. precipitating the wax from the mixture by chilling;
3. recovering the solvent from the wax and dewaxed oil for recycling by distillation and steam stripping.

Two solvents are used. Toluene is used to dissolve the oil and maintain fluidity at low temperatures. Methyl ethyl ketone (MEK) is used because it dissolves little wax at low temperatures and acts as a wax precipitating agent. Other solvents that are sometimes used include benzene, methyl isobutyl ketone, propane, petroleum naphtha, ethylene dichloride, methylene chloride, and sulfur dioxide. In addition, there is a catalytic process used as an alternative to solvent dewaxing.

Solvent treatment systems are designed as closed processes. They generally are operated at low pressures. Despite this there is the potential for fire from a leak or spill contacting a source of ignition such as the drier or extraction heater. In solvent dewaxing, disruption of the vacuum will create a potential fire hazard by allowing air to enter the unit. Extraction solvents that can become fugitive emissions are phenol, furfural, glycols, MEK, amines, and other process chemicals. A simplified process flow sheet for solvent dewaxing is shown in Figure 1.20.

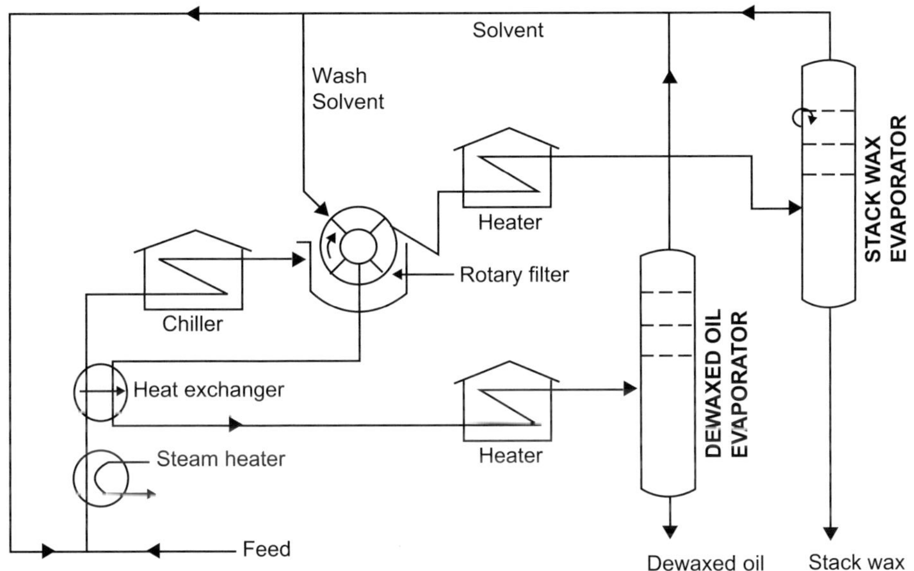

Figure 1.20 Simple schematic of solvent dewaxing.

1.5.6 Thermal cracking

Simple distillation of crude oil produces amounts and types of products that are not consistent with those required by the marketplace. Subsequent refinery processes change the product mix by altering the molecular structure of the hydrocarbons. This is accomplished by the method known as "cracking".

Cracking is a process that breaks or cracks the heavier, higher-boiling-point petroleum fractions into more valuable products such as gasoline, fuel oil, and gas oils. The two basic types of cracking are thermal cracking, using heat and pressure, and catalytic cracking.

The original thermal cracking process was developed circa 1913. Distillate fuels and heavy oils were heated under pressure in large drums until they cracked into smaller molecules with better antiknock characteristics. However, this method produced large amounts of solid coke for which there was no market in the early days. The early process has evolved into the following applications of thermal cracking: visbreaking, steam cracking, and coking.

Visbreaking process

This is a mild form of thermal cracking. It greatly lowers the viscosity of heavy crude-oil residue without affecting the boiling point range. Residual from the

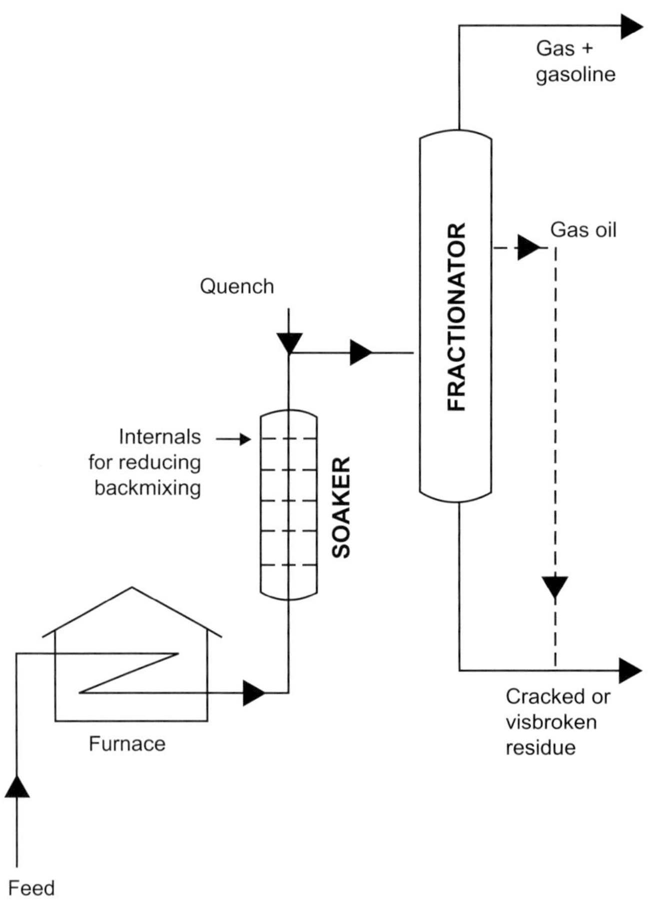

Figure 1.21 Simple schematic of the visbreaking process.

atmospheric distillation tower is heated between 800 and 950°F at atmospheric pressure and mildly cracked in a heater. It is then quenched with cool gas oil to control overcracking, and flashed in a distillation tower.

Visbreaking is used to reduce the pour point of waxy residues and reduce the viscosity of residues used for blending with lighter fuel oils. Middle distillates may also be produced. The thermally cracked residue tar, which accumulates in the bottom of the fractionation tower, is vacuum flashed in a stripper and the distillate is recycled. A simplified process flow sheet for visbreaking is shown in Figure 1.21.

Steam cracking process

This is a petrochemical process used in refineries to produce olefinic raw materials (e.g. ethylene) from various feedstock for petrochemicals manufacture. The feedstock ranges from ethane to vacuum gas oil, with heavier feeds giving higher yields of by-products such as naphtha. Common feeds are ethane, butane, and naphtha.

Steam cracking is carried out at temperatures of 1500–1600°F and at pressures slightly above atmospheric. Naphtha produced from steam cracking contains benzene, which is extracted prior to hydrotreating. Residual from steam cracking is sometimes blended into heavy fuels.

Coking processes

Coking is a severe method of thermal cracking used to upgrade heavy residuals into lighter products or distillates. Coking produces straight-run gasoline known as coker naphtha along with various middle-distillate fractions used as catalytic cracking feedstock. The process so completely reduces hydrogen that the residue is a form of carbon called "coke".

The two most common processes are delayed coking and continuous coking, the latter of which is referred to as contact or fluid coking. Three types of coke are obtained: sponge coke, honeycomb coke, and needle coke. The type of coke produced depends on the reaction mechanism, time, and temperature, and the crude feedstock.

Delayed coking

In this method the heated charge is residuum from atmospheric distillation towers. The residuum is transferred to large coke drums, which provide the long residence time needed to allow the cracking reactions to proceed to completion. Initially the heavy feedstock is fed to a furnace, which heats the residuum to high temperatures (900–950°F) at low pressures of between 25 and 30 psi. The process is controlled to prevent premature coking in the heater tubes. The mixture is passed from the heater to one or more coker drums where the hot material is held for approximately 24 hours; hence the term delayed coking. This interim period is conducted at pressures of 25–75 psi and allows the residuum to undergo cracking into lighter products. Vapors from the drums are returned to a fractionator where gas, naphtha, and gas oils are separated out and recovered. The heavier hydrocarbons produced in the fractionator are recycled through the furnace.

Once the coke reaches a predetermined level in one drum, the flow is diverted to another drum to maintain continuous operation. The full drum is steamed to strip out uncracked hydrocarbons, cooled by water injection, and decoked by mechanical or hydraulic methods. The coke is mechanically removed by an auger rising from the bottom of the drum. Hydraulic decoking consists of fracturing the coke bed with high-pressure water ejected from a rotating cutter. Figure 1.22 shows a simplified process flow sheet for delayed coking.

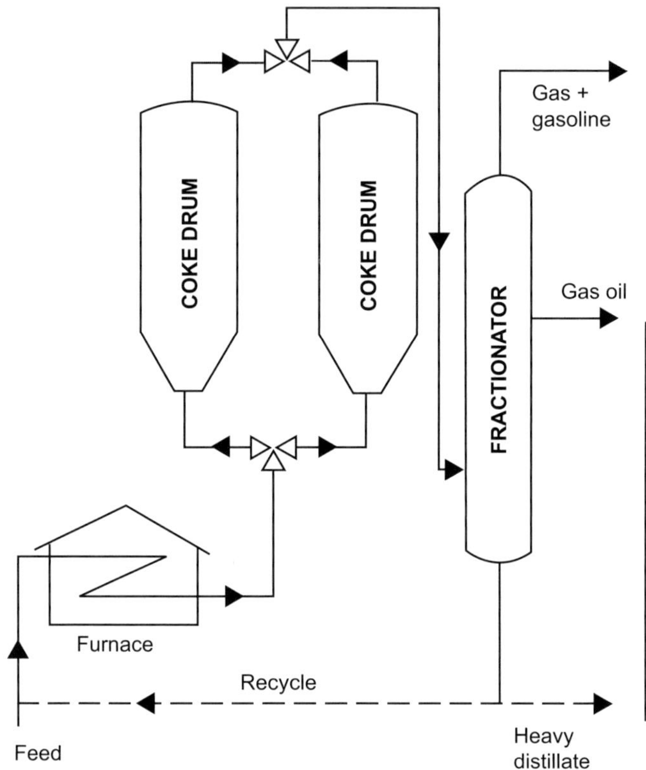

Figure 1.22 Simple schematic of delayed coking.

Continuous coking

The continuous process is known as contact or fluid coking. It consists of a moving-bed process that operates at temperatures higher than with delayed coking. In continuous coking, thermal cracking occurs by using heat transferred from hot, recycled coke particles to feedstock in a radial mixer, called a reactor, at a pressure of 50 psi. Gases and vapors are removed from the reactor, quenched to stop any further reaction, and fractionated. The reacted coke enters a surge drum and is lifted to a feeder and classifier where the larger coke particles are removed as product. The remaining coke is collected in a preheater for recycling with feedstock. Coking occurs both in the reactor and in the surge drum. The process is automatic in that there is a continuous flow of coke and feedstock.

Like other refinery processes thermal cracking is a closed process. The primary potential for fire is from leaks or releases of liquids, gases, or vapors reaching an ignition source such as a heater. The potential for fire is present in coking operations due to vapor or product leaks. Should coking temperatures get out of control, an exothermic reaction can occur within the coker. In thermal cracking when sour crudes are processed, corrosion can occur where metal temperatures are between 450 and 900°F. Above 900°F coke forms a protective layer on the metal. The furnace, soaking drums, lower part of the tower, and high-temperature exchangers are subject to corrosion. Hydrogen sulfide corrosion in coking can also occur when temperatures are not properly controlled above 900°F. Corrosion leads to equipment failures and promotes leaks and fugitive emissions.

Thermal stress is also a great concern in thermal cracking techniques. Continuous thermal changes can lead to bulging and cracking of coke drum shells. In coking, temperature control must often be held within a narrow 10–20°F range, as high temperatures will produce coke that is too hard to cut out of the drum. Low temperatures result in a high asphaltic-content slurry.

Water or steam injection may be used to prevent buildup of coke in delayed coker furnace tubes. Water must be completely drained from the coker, so as not to cause an explosion upon recharging with hot coke.

There are a range of hazardous gases generated from coking operations such as hydrogen sulfide, carbon monoxide, and polynuclear aromatics (PNAs). When coke is handled as a slurry, oxygen depletion can occur within confined spaces such as storage silos (note: wet carbon will adsorb oxygen).

Process wastewater is generally highly alkaline and contains oil, sulfides, ammonia, and phenol. The potential exists in the coking process for exposure to burns when handling hot coke or in the event of a steam-line leak, or from steam, hot water, hot coke, or hot slurry that may be expelled when opening cokers.

1.5.7 *Catalytic cracking*

The purpose of catalytic cracking is to break down complex hydrocarbons into simpler molecules in order to increase the quality and quantity of lighter, more desirable products and to decrease the amount of residuals. Catalytic cracking rearranges the molecular structure of hydrocarbon compounds to convert heavy hydrocarbon feedstock into lighter fractions such as kerosene, gasoline, LPG, heating oil, and petrochemical feedstock. The process is similar to thermal cracking except that catalysts are used to facilitate the conversion of the heavier molecules into lighter products. Use of a catalyst in the cracking reaction increases the yield of improved-quality products under much less severe operating conditions than in thermal cracking. Typical temperatures are 850–950°F and at much lower pressures of 10–20 psi. The catalysts used in refinery cracking units are typically solid materials that include zeolite, aluminum hydrosilicate, treated bentonite clay, fuller's earth, bauxite, and silica–alumina. Catalysts come in the form of powders, beads, pellets, or shaped materials called extrudites. The three basic functions in the catalytic cracking process are as follows:

- Reaction – feedstock reacts with catalyst and cracks into different hydrocarbons.
- Regeneration – catalyst is reactivated by burning off coke.
- Fractionation – cracked hydrocarbon stream is separated into various products.

The types of catalytic cracking processes are:

- fluid catalytic cracking (FCC);
- moving-bed catalytic cracking;
- thermofor catalytic cracking (TCC).

In addition to cracking, catalytic activities include dehydrogenation, hydrogenation, and isomerization.

1.5.8 Fluid catalytic cracking

In the FCC process oil is cracked in the presence of a finely divided catalyst, which is maintained in an aerated or fluidized state by the oil vapors. The fluid cracker consists of a catalyst section and a fractionating section that operate together as an integrated processing unit. The catalyst section contains the reactor and regenerator. The unit is equipped with a standpipe and riser that form the catalyst circulation unit. The fluid catalyst is continuously circulated between the reactor and the regenerator using air, oil vapors, and steam as the conveying media.

The FCC mixes a preheated hydrocarbon charge with hot, regenerated catalyst as it enters the riser leading to the reactor. The charge is combined with a recycle stream within the riser, vaporized, and raised to reactor temperature of 900–1000°F by the hot catalyst. As the mixture travels up the riser, the charge is cracked at a pressure of 10–30 psi.

Modern FCC units allow the cracking to take place in the riser. In these units the reactor serves as a holding vessel whereby the cracking continues until the oil vapors are separated from the catalyst in the reactor cyclones. The resultant product stream or cracked product is then charged to a fractionating column where it is separated into fractions, and some of the heavy oil is recycled to the riser.

Spent catalyst is regenerated to get rid of coke that collects on the catalyst during the process. Spent catalyst flows through the catalyst stripper to the regenerator, where most of the coke deposits burn off at the bottom where preheated air and spent catalyst are mixed. Fresh catalyst is added and spent catalyst is removed to optimize the cracking process. A simplified process flow sheet for fluid catalytic cracking is shown in Figure 1.23.

1.5.9 Moving-bed catalytic cracking

The moving-bed catalytic cracking process is similar to the FCC process. The catalyst is moved continuously to the top of the unit by conveyor or pneumatic lift tubes to a storage hopper, then flows downward by gravity through the reactor, and finally to a regenerator. The regenerator and hopper are isolated from the reactor by steam seals. The cracked product is separated into recycle gas, oil, clarified oil, distillate, naphtha, and wet gas.

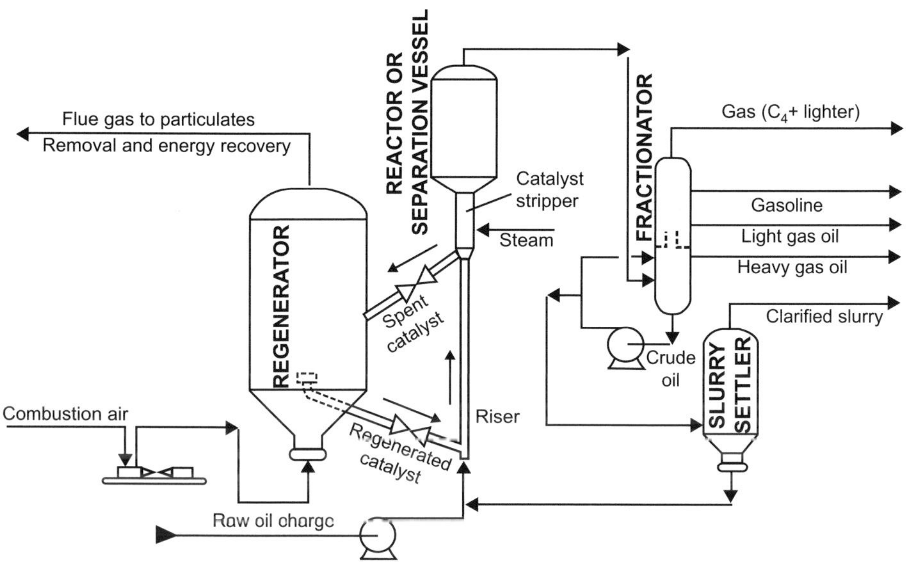

Figure 1.23 Simple schematic of fluid catalytic cracking.

1.5.10 Thermofor catalytic cracking

In a thermofor catalytic cracking unit, the preheated feedstock flows by gravity through the catalytic reactor bed. The vapors are separated from the catalyst and sent to a fractionating tower. The spent catalyst is regenerated, cooled, and recycled. The flue gas from regeneration is sent to a carbon-monoxide boiler for heat recovery.

In all of the above processes the liquid hydrocarbons in the catalyst or entering the heated combustion air stream require monitoring and careful control in order to avoid exothermic reactions from occurring. Because of the presence of heaters in catalytic cracking units, the possibility exists for fire due to a leak or vapor release.

Another concern is that explosive concentrations of catalyst dust during recharge or disposal are possible. When unloading any coked catalyst, the possibility exists for iron sulfide fires. Iron sulfide will ignite spontaneously when exposed to air and therefore must be wetted with water to prevent it from igniting flammable vapors. Coked catalyst may be either cooled below 120°F before it is dumped from the reactor, or dumped into containers that have been purged and inerted with nitrogen and then cooled before further processing. The presence of contaminants or poisons in the process stream is a concern. Regular sampling and testing of the feedstock, product, and recycle streams is required to assure that the cracking process is working as intended and that no contaminants are present. Corrosives or deposits in the feedstock can foul gas compressors.

Inspections of critical equipment including pumps, compressors, furnaces, and heat exchangers are generally required on a frequent basis. Corrosion usually occurs at temperatures below 900°F when refining sour crude. Corrosion occurs where both liquid and vapor phases exist, and at areas subject to local cooling such as nozzles and platform supports.

The processing of high-nitrogen feedstock results in the formation of ammonia and cyanide, which subject carbon steel equipment in the FCC overhead system to corrosion, stress cracking, or hydrogen blistering. Attempts to minimize these actions are done by water wash or corrosion inhibitors. Water wash may also be used to protect overhead condensers in the main column subjected to fouling from ammonium hydrosulfide. Inspections for leaks due to erosion or other malfunctions such as catalyst buildup on the expanders, coking in the overhead feeder lines from feedstock residues, and other unusual operating conditions must be done on a frequent and aggressive basis. Hydrogen sulfide and/or carbon monoxide will be released with spilled product or as fugitive vapors.

Catalyst regeneration involves steam stripping and decoking. This process produces fluid waste streams that contain varying amounts of hydrocarbons, phenol, ammonia, hydrogen sulfide, mercaptan, and other materials depending upon the feedstock, crudes, and processes. Inadvertent formation of nickel carbonyl can occur in cracking processes using nickel catalysts, with resultant potential for hazardous exposures and releases.

1.5.11 Hydrocracking

Hydrocracking is a two-stage process that combines catalytic cracking and hydrogenation. The heavier feedstocks are cracked in the presence of hydrogen to produce more desirable products. The process employs high pressure, high temperature, a catalyst, and hydrogen. The process is used for feedstocks that are difficult to process by either catalytic cracking or reforming. The feedstocks are characterized usually by a high polycyclic aromatic content and/or high concentrations of the catalyst poisons, sulfur and nitrogen compounds. Figure 1.24 shows a simplified process flow sheet for two-stage hydrocracking.

The hydrocracking process depends on the nature of the feedstock and the relative rates of the two competing reactions, hydrogenation and cracking. Heavy aromatic feedstock is converted into lighter products under very high pressures (1000–2000 psi) and high temperatures of 750–1500°F in the presence of hydrogen and special catalysts. When the feedstock has a high paraffinic content, the primary function of hydrogen is to prevent the formation of polycyclic aromatic compounds (PAHs). Another important role of hydrogen is to reduce tar formation and prevent buildup of coke on the catalyst. Hydrogenation also serves to convert sulfur and nitrogen compounds present in the feedstock to hydrogen sulfide and ammonia. The process produces large amounts of isobutane for alkylation feedstock. Hydrocracking also performs isomerization

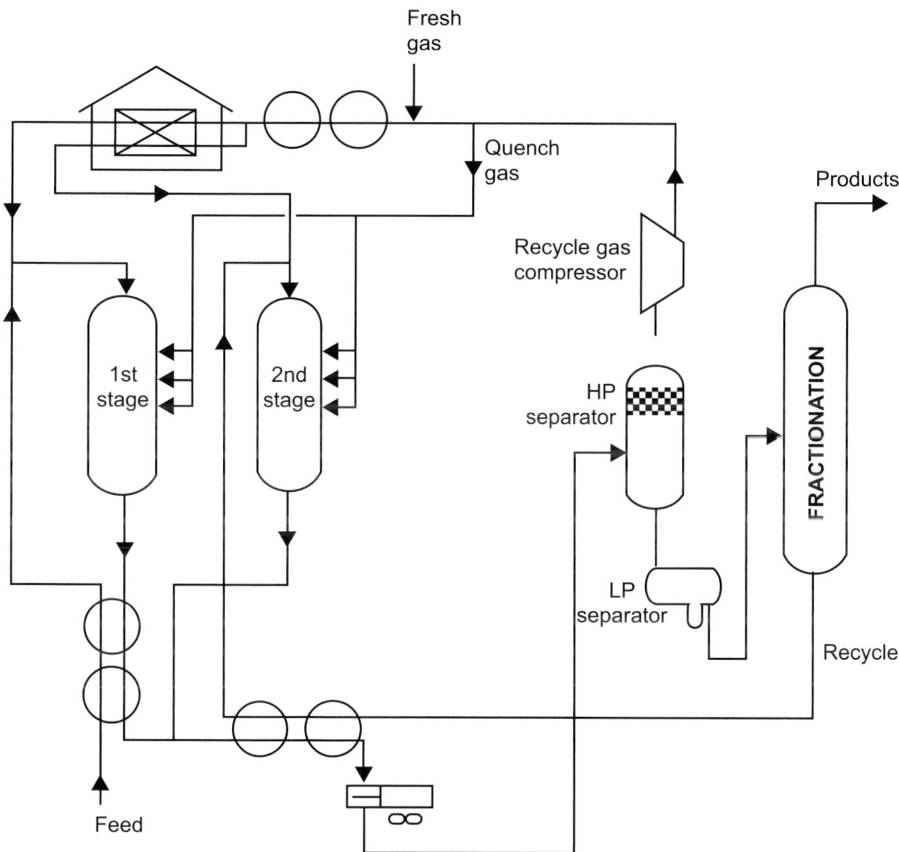

Figure 1.24 Simple schematic of two-stage hydrocracking.

for pour-point control and smoke-point control. These properties are important in high-quality jet fuel.

In the first stage of the hydrocracking process, preheated feedstock is mixed with recycled hydrogen and sent to the first-stage reactor, where catalysts convert sulfur and nitrogen compounds to hydrogen sulfide and ammonia. Limited hydrocracking also occurs.

After the hydrocarbon leaves the first stage, it is cooled and liquefied and then sent through a hydrocarbon separator. Hydrogen is recycled to the feedstock. The liquid is charged to a fractionator. Depending on the products desired (gasoline components, jet fuel, and gas oil), the fractionator is run to cut out some portion of the first-stage reactor out-turn. Kerosene-range material can be taken as a separate side-draw product or included in the fractionator bottoms with the gas oil.

Fractionator bottoms are mixed with a hydrogen stream and charged to the second stage. Since this material has already been subjected to some hydrogenation, cracking, and reforming in the first stage, the operations of the second stage are more severe with higher temperatures and pressures. As with the first stage, the second-stage product is separated from the hydrogen and charged to the fractionator.

The high pressures and temperature of the unit mandate careful control of both hydrocarbon leaks and hydrogen releases to prevent fires. Care is also needed to ensure that explosive concentrations of catalytic dust do not form during the recharging stage.

Frequent inspection and testing of safety relief devices are important due to the very high pressures in this unit. Proper process control is needed to protect against plugging reactor beds. Unloading coked catalyst requires special precautions to prevent iron sulfide-induced fires. The coked catalyst should either be cooled to below 120°F before dumping, or be placed in nitrogen-inerted containers until cooled. The hydrogen sulfide content of the feedstock must be strictly controlled to a minimum to reduce the possibility of severe corrosion under the high operating pressures and temperatures of the unit. Corrosion by wet carbon dioxide in areas of condensation also must be considered. When processing high-nitrogen feedstock, the ammonia and hydrogen sulfide that form ammonium hydrosulfide are responsible for severe corrosion at temperatures below the water dew point. Ammonium hydrosulfide is also present in sour-water stripping.

There is always a potential for exposure to hydrocarbon gas and to vapor emissions, hydrogen and hydrogen sulfide gas from high-pressure leaks. Large quantities of carbon monoxide may be released during catalyst regeneration and changeover. Catalyst steam stripping and regeneration create waste streams containing sour water and ammonia.

1.5.12 Catalytic reforming

Catalytic reforming is a process used to convert low-octane naphthas into high-octane gasoline blending components called reformates. Reforming is the total effect of several reactions that occur simultaneously including cracking, polymerization, dehydrogenation, and isomerization. Depending on the properties of the naphtha feedstock (as measured by the paraffin, olefin, naphthene, and aromatic content) and catalysts used, reformates can be produced with high concentrations of toluene, benzene, xylene, and other aromatics useful in gasoline blending and petrochemical processing. Hydrogen, a significant by-product, is separated from the reformate for recycling and use in other processes. A catalytic reformer consists of a reactor section and a product-recovery section. A feed preparation section consists of a combination of hydrotreatment and distillation, which allows the feedstock to be prepared to specification. Most processes use platinum as the active catalyst. Sometimes platinum is combined with a second catalyst (bimetallic catalyst) such as rhenium or another noble metal. There are a number of different commercial catalytic reforming processes. Principal ones are plat-forming, powerforming, ultraforming, and thermofor catalytic reforming. In the platforming process, the first step is the preparation of the naphtha feed to remove impurities and reduce catalyst degradation. The naphtha feedstock is then mixed with hydrogen, vaporized, and passed through a series of alternating furnace and fixed-bed reactors containing the catalyst. The effluent from the last reactor is cooled and sent to a separator to remove the hydrogen-rich gas stream from the top of the separator for recycling. The liquid product from the bottom of the separator is sent to a fractionator called a stabilizer (or butanizer). The butanizer makes a bottom product called reformate; butanes and lighter go overhead and are sent to the saturated gas plant. Figure 1.25 shows a simplified process flow sheet for catalytic reforming.

Some catalytic reformers operate at low pressure in the range of 50–200 psi, and others operate at high pressures up to 1000 psi. Some catalytic reforming systems continuously regenerate the catalyst in other systems. One reactor at a time is taken off-stream for catalyst regeneration, and some facilities regenerate all of the reactors during turnarounds.

The potential for fire exists should a leak or release of reformate gas or hydrogen occur. Operating procedures are aimed in part at the control of hot spots during startup. Safe catalyst handling is very important. Care must be taken not to break or crush the catalyst when loading the beds, as the small fines

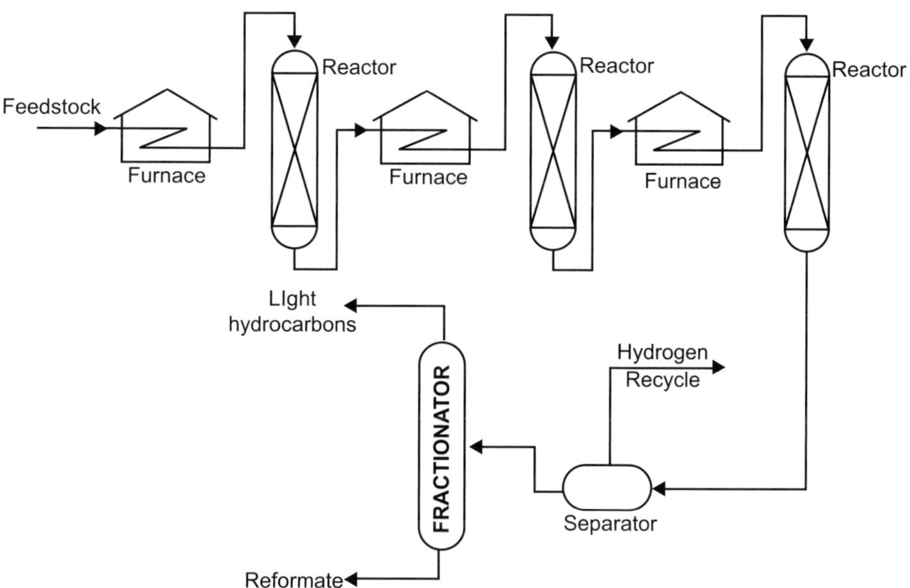

Figure 1.25 Simple schematic of platforming process.

will plug up the reformer screens. Precautions against dust when regenerating or replacing catalyst are also considered. Water wash is used where stabilizer fouling is suspected due to the formation of ammonium chloride and iron salts. Ammonium chloride may form in pretreater exchangers and cause corrosion and fouling. Hydrogen chloride from the hydrogenation of chlorine compounds forms acid or ammonium chloride salt. There is potential for exposure to hydrogen sulfide and benzene emissions from a leak or release. Also fugitive emissions of carbon monoxide and hydrogen sulfide may occur during regeneration of catalyst.

1.5.13   *Catalytic hydrotreating*

Catalytic hydrotreating is a hydrogenation process used to remove about 90% of contaminants such as nitrogen, sulfur, oxygen, and metals from liquid petroleum fractions. These contaminants can have detrimental effects on the equipment, the catalysts, and the quality of the finished product. Hydrotreating is done prior to processes such as catalytic reforming so that the catalyst is not contaminated by untreated feedstock. Hydrotreating is also used prior to catalytic cracking to reduce sulfur and improve product yields, and to upgrade middle-distillate petroleum fractions into finished kerosene, diesel fuel, and heating fuel oils. The process also converts olefins and aromatics to saturated compounds.

Hydrotreating for sulfur removal is called hydrodesulfurization. In this process the feedstock is deaerated and mixed with hydrogen, preheated in a fired heater (600–800°F), and then charged under pressure (up to 1000 psi) through a fixed-bed catalytic reactor. In the reactor, the sulfur and nitrogen compounds in the feedstock are converted into H_2S and NH_3. The reaction products leave the reactor and after cooling to a low temperature enter a liquid/gas separator.

The hydrogen-rich gas from the high-pressure separation is recycled to combine with the feedstock, and the low-pressure gas stream rich in H_2S is sent to a gas-treating unit where H_2S is removed. The clean gas is then suitable as fuel for the refinery furnaces. The liquid stream is the product from hydrotreating and is normally sent to a stripping column for removal of H_2S and other undesirable components. In cases where steam is used for stripping, the product is sent to a vacuum drier for removal of water. Hydrodesulfurized products are blended or used as catalytic reforming feedstock. A simplified process flow diagram for hydrodesulfurization is shown in Figure 1.26.

There are a range of other hydrotreating processes aimed at different products. The processes differ depending upon the feedstock available and catalysts used. Hydrotreating can be used to improve the burning characteristics of distillates such as kerosene. Hydrotreatment of a kerosene fraction can convert aromatics into naphthenes, which are cleaner-burning compounds.

Lube-oil hydrotreating employs catalytic treatment of the oil with hydrogen to improve product quality. The objectives in mild lube hydrotreating include

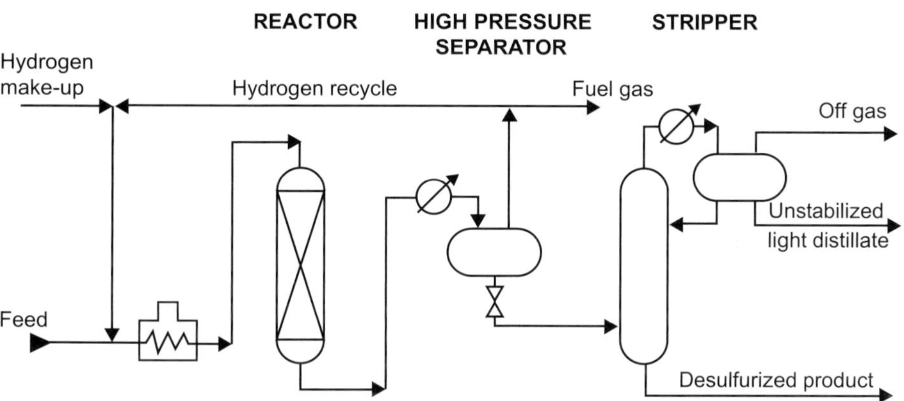

Figure 1.26 Simple schematic of distillate hydrodesulfurization.

saturation of olefins and improvements in color, odor, and acid nature of the oil. Mild lube hydrotreating also may be used following solvent processing. Operating temperatures are below 600°F and operating pressures below 800 psi. Severe lube hydrotreating is performed at temperatures around 600–750°F and hydrogen pressures up to 3000 psi. These severe operations are capable of saturating aromatic rings, along with sulfur and nitrogen removal, imparting specific properties not achieved at mild conditions.

Hydrotreating is also used to improve the quality of pyrolysis gasoline (pygas), a by-product from the manufacture of ethylene. The outlet for pygas has been motor gasoline blending, a suitable route in view of its high octane number. Only small portions can be blended untreated owing to the unacceptable odor, color, and gum-forming tendencies of this material. The quality of pygas, which is high in diolefin content, can be satisfactorily improved by hydrotreating, whereby conversion of diolefins into mono-olefins provides an acceptable product for motor gas blending.

A primary concern with any of these processes is the potential for fire in the event of a leak or release of product or hydrogen gas. Most processes require hydrogen generation to provide for a continuous supply. Because of the operating temperatures and presence of hydrogen, the hydrogen sulfide content of the feedstock must be strictly controlled to a minimum to reduce corrosion. Hydrogen chloride may form and condense as hydrochloric acid in the lower-temperature parts of the unit. Ammonium hydrosulfide may form in high-temperature, high-pressure units. Excessive contact time and/or temperature will cause coking. Precautions need to be taken when unloading coked catalyst from the unit to prevent iron sulfide fires. The coked catalyst must be cooled to below 120°F before removal, or dumped into nitrogen-inerted bins where it can be cooled before further handling. Special antifoam additives are used to prevent catalyst poisoning from silicone carryover in the coker feedstock. There is a potential for release of hydrogen sulfide or hydrogen gas, or ammonia should a sour-water leak or spill occur. Phenol also may be present if high-boiling-point feedstocks are processed.

1.5.14 Isomerization

Isomerization converts n-butane, n-pentane, and n-hexane into their respective isoparaffins of substantially higher octane number. The straight-chain paraffins are converted to their branched-chain counterparts whose component atoms are the same but are arranged in a different geometric structure.

Isomerization is important for the conversion of n-butane into isobutane, to provide additional feedstock for alkylation units, and the conversion of normal pentanes and hexanes into higher branched isomers for gasoline blending. Isomerization is similar to catalytic reforming in that the hydrocarbon molecules are rearranged, but, unlike catalytic reforming, isomerization just converts normal paraffins to isoparaffins. There are two isomerization processes, butane (C_4) and pentane/hexane (C_5/C_6). Butane isomerization produces feedstock for alkylation. Aluminum chloride catalyst plus hydrogen chloride are used for the low-temperature processes. Platinum or another metal catalyst is used for the higher-temperature processes. In a typical low-temperature process, the feed to the isomerization plant is n-butane or mixed butanes mixed with hydrogen to inhibit olefin formation. The feed is passed to the reactor at 230–340°F and 200–300 psi. Hydrogen is flashed off in a high-pressure separator and the hydrogen chloride removed in a stripper column. The resultant butane mixture is sent to a fractionator (called a deisobutanizer) to separate n-butane from the isobutane product. Pentane/hexane isomerization increases the octane number of the light gasoline components n-pentane and n-hexane, which are found in abundance in straight-run gasoline. In a typical C_5/C_6 isomerization process, dried and desulfurized feedstock is mixed with a small amount of organic chloride and recycled hydrogen, and then heated to reactor temperature. It is passed over supported-metal catalyst in the first reactor, where benzene and olefins are hydrogenated. The feed then goes to the isomerization reactor, where the paraffins are catalytically isomerized to isoparaffins. The reactor effluent is then cooled and subsequently separated in the product separator into two streams: a liquid product (isomerate) and a recycle hydrogen-gas stream. The isomerate is washed with caustic and water, acid stripped, and stabilized before going to storage.

Problems occur when the feedstock is not completely dried and desulfurized as the potential exists for acid formation leading to catalyst poisoning and metal corrosion. Water or steam must not be allowed to enter areas where hydrogen chloride is present. Precautions are needed to prevent HCl from entering sewers and drains. There is a potential for exposure to hydrogen gas, hydrochloric acid and hydrogen chloride, and to dust when the solid catalyst is used. Figures 1.27 and 1.28 show process flow sheets for isomerization.

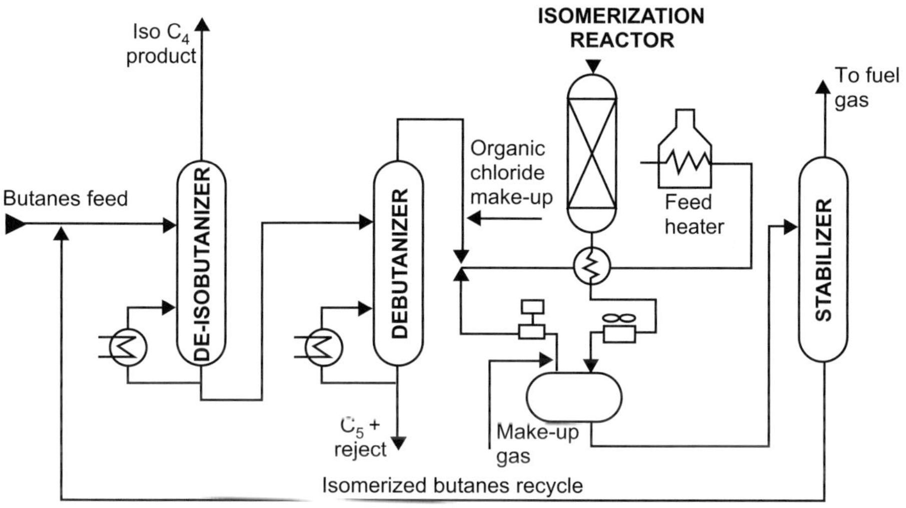

Figure 1.27 Simple schematic of C$_4$ isomerization.

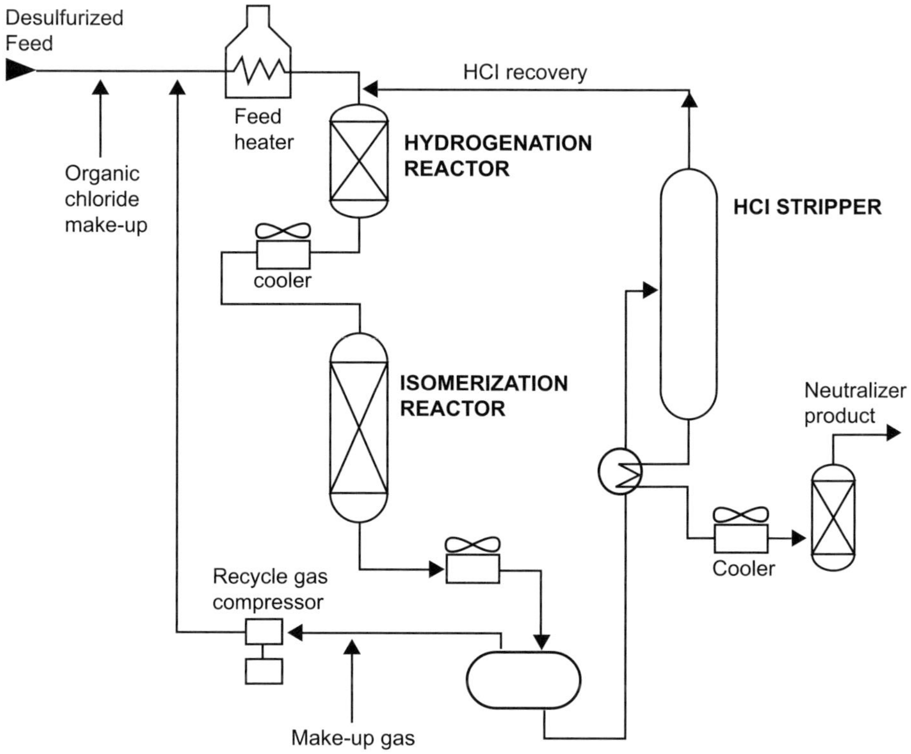

Figure 1.28 Simple schematic of C$_5$ and C$_6$ isomerization.

1.5.15 *Polymerization*

Polymerization converts light olefin gases including ethylene, propylene, and butylene into hydrocarbons of higher molecular weight and higher octane number. These products are used as gasoline blending stocks. Polymerization combines two or more identical olefin molecules to form a single molecule with the same elements in the same proportions as in the original molecules. Polymerization may be accomplished thermally or in the presence of a catalyst at lower temperatures.

The olefin feedstock is pretreated to remove sulfur and other undesirable compounds. In the catalytic process the feedstock is either passed over a solid phosphoric acid catalyst or comes in contact with liquid phosphoric acid, where an exothermic polymeric reaction takes place. The reaction requires cooling water and the injection of cold feedstock into the reactor to control temperatures between 300 and 450°F at pressures from 200 to 1200 psi. The reaction products leaving the reactor are sent to stabilization and/or fractionator systems to separate saturated and unreacted gases from the polymer gasoline product.

Polymerization is a closed process where the potential for a fire exists due to leaks or releases reaching a source of ignition. The potential for an uncontrolled exothermic reaction exists should loss of cooling water occur. Severe corrosion leading to equipment failure will occur should water make contact with the phosphoric acid, such as during water washing at shutdowns. Corrosion may occur in piping manifolds, reboilers, exchangers, and other locations where acid may settle out. Because this is a closed system, exposures are expected to be minimal under normal operating conditions. There is a potential for exposure to caustic wash (sodium hydroxide), to phosphoric acid used in the process or washed out during turnarounds, and to catalyst dust. Figure 1.29 shows a simplified process flow diagram for the polymerization process.

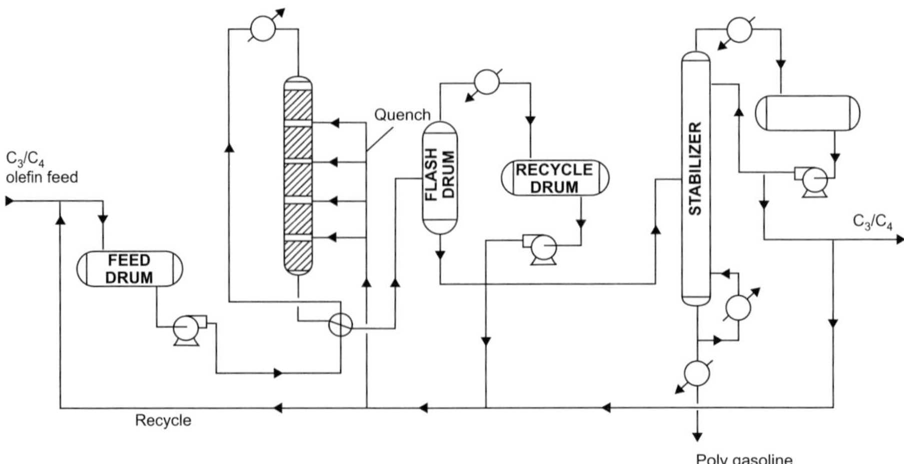

Figure 1.29 Simplified process flow diagram for polymerization.

1.5.16 Alkylation

Alkylation is a process which combines low-molecular-weight olefins (primarily a mixture of propylene and butylene) with isobutene in the presence of a catalyst, either sulfuric acid or hydrofluoric acid. The product is referred to as alkylate and is composed of a mixture of high-octane, branched-chain paraffinic hydrocarbons.

Alkylate is a premium blending stock because it has exceptional antiknock properties and is clean burning. The octane number of the alkylate depends mainly upon the kinds of olefin used and upon operating conditions.

In cascade-type sulfuric acid (H_2SO_4) alkylation units, the feedstock (propylene, butylene, amylene, and fresh isobutane) enters the reactor and contacts the concentrated sulfuric acid catalyst. The reactor is divided into zones, with olefins fed through distributors to each zone, and the sulfuric acid and isobutanes flowing over baffles from zone to zone. The reactor effluent is separated into hydrocarbon and acid phases in a settler, and the acid is returned to the reactor. The hydrocarbon phase is hot-water washed with caustic for pH control before being successively depropanized, deisobutanized, and debutanized. The alkylate obtained from the deisobutanizer can then go directly to motor-fuel blending or be rerun to produce aviation-grade blending stock. The isobutane is recycled to the feed.

Two hydrofluoric acid alkylation schemes are the Phillips and UOP processes. In the Phillips process, olefin and isobutane feedstock are dried and fed to a combination reactor/settler system. Upon leaving the reaction zone, the reactor effluent flows to a settler (separating vessel) where the acid separates from the hydrocarbons. The acid layer at the bottom of the separating vessel is recycled. The hydrocarbon phase, consisting of propane, normal butane, alkylate, and excess (recycle) isobutane, is charged to the main fractionators. This is the upper phase. The bottom product is motor alkylate. The main fractionator overhead, consisting mainly of propane, isobutane, and HF, goes to a depropanizer. Propane with trace amounts of HF goes to an HF stripper for HF removal and is then catalytically defluorinated, treated, and sent to storage. Isobutane is withdrawn from the main fractionator and recycled to the reactor/settler, and alkylate from the bottom of the main fractionator is sent to product blending.

The UOP process uses two reactors with separate settlers. Half of the dried feedstock is charged to the first reactor, along with recycle and makeup isobutane. The reactor effluent then goes to its settler, where the acid is recycled and the hydrocarbon is charged to the second reactor. The other half of the feedstock also goes to the second reactor, with the settler acid being recycled and the hydrocarbons charged to the main fractionator. Subsequent processing is similar to the Phillips process. Overhead from the main fractionator goes to a depropanizer. Isobutane is recycled to the reaction zone and alkylate is sent to product blending. Both processes are illustrated in Figure 1.30.

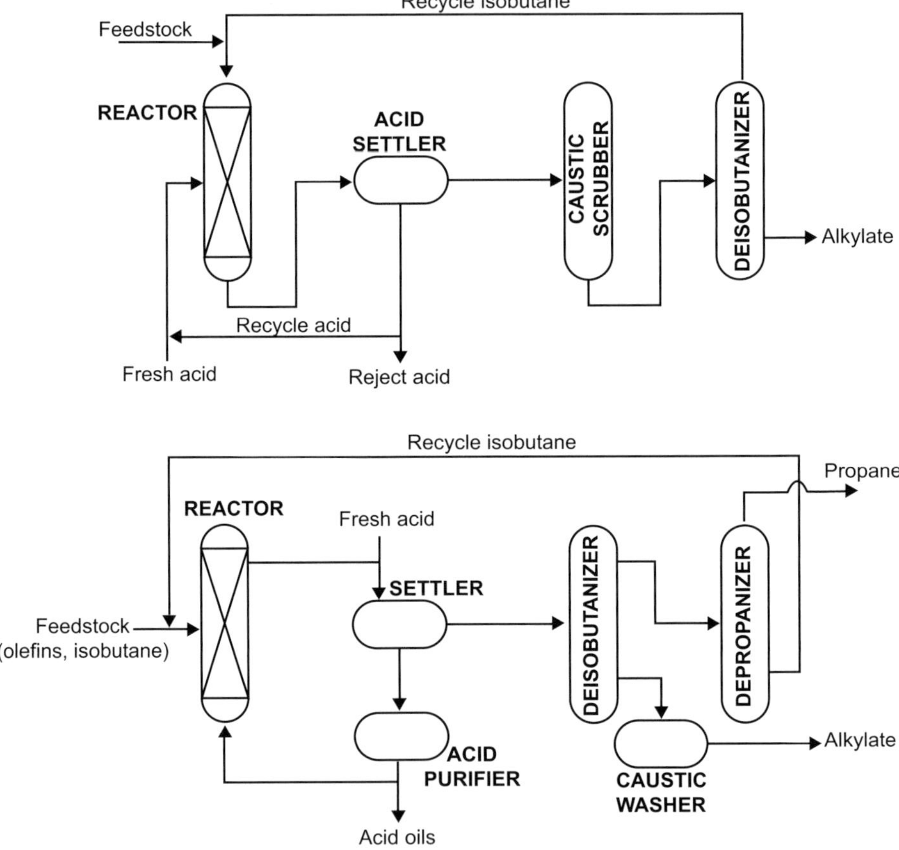

Figure 1.30 Common alkylation process schemes, Phillips (upper) and UOP (lower).

Leaks, spills, or releases involving hydrofluoric acid or hydrocarbons containing hydrofluoric acid can be extremely hazardous. Process unit containment by curbs, drainage, and isolation so that effluent can be neutralized before release to the sewer system is considered. Vents can be routed to soda-ash scrubbers to neutralize hydrogen fluoride gas or hydrofluoric acid vapors before release. Pressure on the cooling water and steam side of exchangers should be kept below the minimum pressure on the acid service side to prevent water contamination. Some corrosion and fouling in sulfuric acid units may occur from the breakdown of sulfuric acid esters or where caustic is added for neutralization. These esters can be removed by fresh acid treating and hot-water washing. To prevent corrosion from hydrofluoric acid, the acid concentration inside the process unit should be maintained above 65% and moisture below 4%.

Because this is a closed process, exposures are expected to be minimal during normal operations. There is a potential for exposure should leaks, spills, or releases occur. Sulfuric acid and (particularly) hydrofluoric acid are potentially hazardous chemicals. Special precautionary emergency preparedness measures and protection appropriate to the potential hazard and areas possibly affected need to be provided. Safe work practices and appropriate skin and respiratory personal protective equipment are needed for potential exposures to hydrofluoric and sulfuric acids during normal operations such as reading gauges, inspecting, and process sampling, as well as during emergency response, maintenance, and turnaround activities.

1.5.17 Sweetening and treating

Treating is a process in which contaminants comprising organic compounds containing sulfur, nitrogen, and oxygen, dissolved metals and inorganic salts, and soluble salts dissolved in emulsified water are removed from petroleum fractions or streams. There are several different treating processes. The primary purpose of the majority of them is the elimination of unwanted sulfur compounds. A variety of intermediate and finished products, including middle distillates, gasoline, kerosene, jet fuel, and sour gases, are dried and sweetened. Sweetening is a major refinery treatment step of gasoline. The process treats sulfur compounds (hydrogen sulfide, thiophene, and mercaptan) to improve color, odor, and oxidation stability. Sweetening also reduces concentrations of carbon dioxide.

Treating can be accomplished at an intermediate stage in the refining process, or just before sending the finished product to storage. Choices of a treating method depend on the nature of the petroleum fractions, amount and type of impurities in the fractions to be treated, the extent to which the process removes the impurities, and end-product specifications. Treating materials include acids, solvents, alkalis, and oxidizing and adsorption agents.

Sulfuric acid is the most commonly used acid treating process. Sulfuric acid treating results in partial or complete removal of unsaturated hydrocarbons, sulfur, nitrogen, and oxygen compounds, and resinous and asphaltic compounds. It is used to improve the odor, color, stability, carbon residue, and other properties of the oil. Clay/lime treatment of acid-refined oil removes traces of asphaltic materials and other compounds, improving product color, odor, and stability. Caustic treating with sodium (or potassium) hydroxide is used to improve odor and color by removing organic acids (naphthenic acids, phenols) and sulfur compounds (mercaptans, H_2S) by a caustic wash. By combining caustic soda solution with various solubility promoters (e.g. methyl alcohol and cresols), up to 99% of all mercaptans as well as oxygen and nitrogen compounds can be dissolved from petroleum fractions.

Feedstocks from various refinery units are sent to gas treating plants where butanes and butenes are removed for use as alkylation feedstock, heavier components are sent to gasoline blending, propane is recovered for

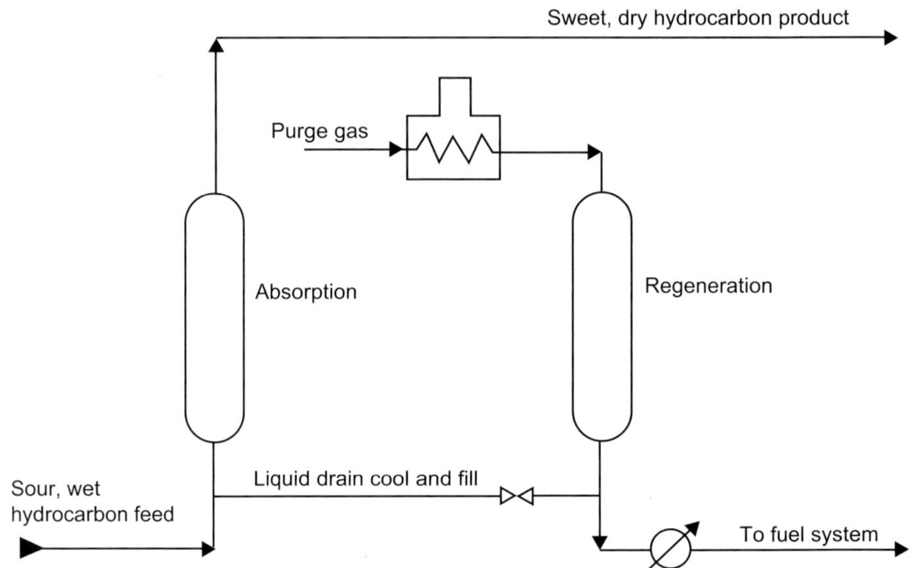

Figure 1.31 Molecular sieve drying and sweetening.

LPG, and propylene is removed for use in petrochemicals. Some mercaptans are removed by water-soluble chemicals that react with the mercaptans. Caustic liquid (sodium hydroxide), amine compounds (diethanolamine) or fixed-bed catalyst sweetening also may be used. Drying is accomplished by the use of water absorption or adsorption agents to remove water from the products. Some processes simultaneously dry and sweeten by adsorption on molecular sieves. Molecular sieve drying and sweetening is illustrated in Figure 1.31.

Sulfur recovery converts hydrogen sulfide in sour gases and hydrocarbon streams to elemental sulfur. The most widely used recovery system is the Claus process, which uses both thermal and catalytic-conversion reactions. A typical process produces elemental sulfur by burning hydrogen sulfide under controlled conditions. Knockout pots are used to remove water and hydrocarbons from feed gas streams. The gases are then exposed to a catalyst to recover additional sulfur. Sulfur vapor from burning and conversion is condensed and recovered.

Current methods of recovering sulfur from H_2S gas streams typically combine two processes: the Claus process, which can be followed by a tail-gas treatment unit (TGTU) for very high sulfur recovery efficiencies. Since the Claus process by itself cannot achieve a high removal efficiency, a TGTU is used to further recover additional sulfur to obtain an overall sulfur removal efficiency of 99.7%.

The Claus process can receive H_2S-rich gas streams from the amine units and the sour-water stripping system. Other components that enter the sulfur recovery unit include ammonia (NH_3), CO_2 and, to a minor extent, various hydrocarbons. The Claus process consists of partial combustion of H_2S-rich gas and reacting the resulting SO_2 and unburned H_2S in the presence of an activated alumina catalyst to produce elemental sulfur. Elemental sulfur is produced according to the following reaction:

$$2H_2S + SO_2 \rightarrow 2H_2O + 3S$$

The Claus process is in the public domain and may be applied at any refinery. A two-stage Claus process is the most common in Europe. There are many licensors of this process. The primary ones are as follows.

In the USA:

- Parsons
- TPA
- Black & Veatch
- Jacobs
- KBR.

Elsewhere:

- Lurgi
- KTI (Technip).

More than a dozen processes for TGTU have been developed to enhance the recovery of sulfur compounds from refinery streams. These include:

- Wet sulfuric acid process, developed by Haldor Topsoe
- Direct sulfur recovery process, developed by Research Triangle Institute
- Bio-FGB process, developed by Paques
- Parson high-activity process
- Amoco's cold-bed absorption process
- Delta Hudson (now Jacobs) process (MCRC™)
- IFP tail-gas treatment (Clauspol)
- Shell Claus off-gas treatment (SCOT)
- Lo-Cat® process, licensed by Merichem
- Marathon Oil's Hysulf process
- Recycle selectrox process, developed by Parsons and Unocal and licensed through UOP.

Other processes are:

- Sulfreen™
- Hydrosulfreen™
- Superclaus™
- Cansolv®
- Clintox
- Z-Sorb.

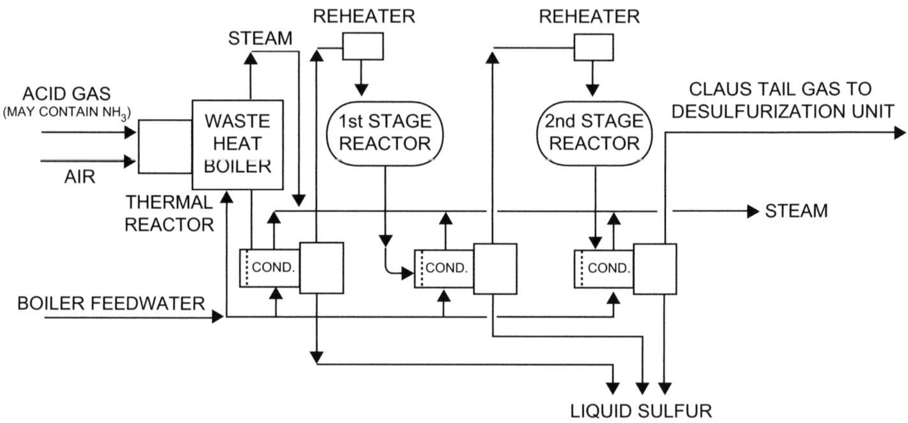

Figure 1.32 Simplified Claus sulfur recovery flow schematic.
Source: US EPA Office of General Enforcement, Petroleum Refinery Enforcement Manual, 1980.

The potential exists for fire from a leak or release of feedstock or product. Sweetening processes use air or oxygen. If excess oxygen enters these processes, it is possible for a fire to occur in the settler due to the generation of static electricity, which acts as the ignition source. Because these are closed processes, exposures are expected to be minimal under normal operating conditions. There is a potential for exposure to hydrogen sulfide, caustic (sodium hydroxide), spent caustic, spent catalyst (Merox), catalyst dust, and sweetening agents (sodium carbonate and sodium bicarbonate). Figure 1.32 provides a simplified schematic of the basic Claus process.

1.5.18 Gas plants

Unsaturated (unsat) gas plants recover light hydrocarbons (C_3 and C_4 olefins) from wet gas receivers. In a typical unsat gas plant, the gases are compressed and treated with amine to remove hydrogen sulfide either before or after they are sent to a fractionating absorber, where they are mixed into a concurrent flow of debutanized gasoline. The light fractions are separated by heat in a reboiler, the off-gas is sent to a sponge absorber, and the bottoms are sent to a debutanizer. A portion of the debutanized hydrocarbon is recycled, with the balance sent to the splitter for separation. The overhead gases go to a depropanizer for use as alkylation unit feedstock. The potential for a fire exists should spills, releases, or vapors reach a source of ignition. In unsat gas plants handling FCC feedstock, the potential exists for corrosion from moist hydrogen sulfide and cyanides. When feedstocks are from the delayed coker or the TCC, corrosion from hydrogen sulfide and deposits in the high-pressure sections of gas compressors from ammonium compounds is possible.

Because these are closed processes, exposures are expected to be minimal under normal operating conditions. There is a potential for exposures to amine compounds such as monoethanolamine (MEA), diethanolamine (DEA), methyldiethanolamine (MDEA), and hydrocarbons.

Saturated (sat) gas plants separate refinery gas components including butanes for alkylation, pentanes for gasoline blending, LPGs for fuel, and ethane for petrochemicals. Because sat gas processes depend on the feedstock and product demand, each refinery uses different systems, usually absorption–fractionation or straight fractionation. In absorption–fractionation, gases and liquids from various refinery units are fed to an absorber–de-ethanizer, where C_2 and lighter fractions are separated from heavier fractions by lean oil absorption and removed for use as fuel gas or petrochemical feed. The heavier fractions are stripped and sent to a debutanizer, and the lean oil is recycled back to the absorber–de-ethanizer. C_3/C_4 is separated from pentanes in the debutanizer, scrubbed to remove hydrogen sulfide, and fed to a splitter where propane and butane are separated. In fractionation sat gas plants, the absorption stage is eliminated.

There is potential for fire if a leak or release reaches a source of ignition such as the unit reboiler. Corrosion could occur from the presence of hydrogen sulfide, carbon dioxide, and other compounds as a result of prior treating. Streams containing ammonia should be dried before processing. Antifouling additives may be used in absorption oil to protect heat exchangers. Corrosion inhibitors may be used to control corrosion in overhead systems. Because this is a closed process, exposures are expected to be minimal during normal operations. There is potential for exposure to hydrogen sulfide, carbon dioxide, and other products such as diethanolamine or sodium hydroxide carried over from prior treating.

1.5.19 Hydrogen production

High-purity hydrogen (95–99%) is required for hydrodesulfurization, hydrogenation, hydrocracking, and petrochemical processes. Hydrogen, produced as a by-product of refinery processes (principally hydrogen recovery from catalytic reformer product gases), often is not enough to meet the total refinery requirements, necessitating the manufacture of additional hydrogen or obtaining supply from external sources.

In steam–methane reforming, desulfurized gases are mixed with superheated steam (1100–1600°F) and reformed in tubes containing a nickel base catalyst. The reformed gas, which consists of steam, hydrogen, carbon monoxide, and carbon dioxide, is cooled and passed through converters containing an iron catalyst, where the carbon monoxide reacts with steam to form carbon dioxide and more hydrogen. The carbon dioxide is removed by amine washing. Any remaining carbon monoxide in the product stream is converted to methane.

Steam–naphtha reforming is a continuous process for the production of hydrogen from liquid hydrocarbons and is, in fact, similar to steam–methane reforming. A variety of naphthas in the gasoline boiling range may be employed, including fuel containing up to 35% aromatics. Following pretreatment to remove sulfur compounds, the feedstock is mixed with steam and taken to the reforming furnace (1250–1500°F), where hydrogen is produced.

The possibility of fire exists should a leak or release occur and reach an ignition source. The potential exists for burns from hot gases and superheated steam should a release occur. Inspections and testing should be considered where the possibility exists for valve failure due to contaminants in the hydrogen. Carryover from caustic scrubbers should be controlled to prevent corrosion in preheaters. Chlorides from the feedstock or steam system should be prevented from entering reformer tubes and contaminating the catalyst. Because these are closed processes, exposures are expected to be minimal during normal operating conditions. There is a potential for exposure to excess hydrogen, carbon monoxide, and/or carbon dioxide. Condensate can be contaminated by process materials such as caustics and amine compounds, with resultant exposures.

1.5.20 Amine plants

Amine plants remove acid contaminants from sour gas and hydrocarbon streams. In amine plants, gas and liquid hydrocarbon streams containing carbon dioxide and/or hydrogen sulfide are charged to a gas absorption tower or liquid contactor, where the acid contaminants are absorbed by counter-flowing amine solutions (i.e. MEA, DEA, MDEA). The stripped gas or liquid is removed overhead, and the amine is sent to a regenerator. In the regenerator, the acidic components are stripped by heat and reboiling action and disposed of, and the amine is recycled. The potential for fire exists where a spill or leak could reach a source of ignition. To minimize corrosion, proper operating practices should be established and regenerator bottom and reboiler temperatures controlled. Oxygen should be kept out of the system to prevent amine oxidation. Because this is a closed process, exposures are expected to be minimal during normal operations. There is potential for exposure to amine compounds (i.e. MEA, DEA, MDEA), hydrogen sulfide, and carbon dioxide.

1.5.21 Asphalt production

Asphalt is a portion of the residual fraction that remains after primary distillation operations. It is further processed to impart characteristics required by its final use. In vacuum distillation, generally used to produce road-tar asphalt, the residual is heated to about 750°F and charged to a column where vacuum is applied to prevent cracking.

Asphalt for roofing materials is produced by air blowing. Residual is heated in a pipe still almost to its flash point and charged to a blowing tower where hot air is injected for a predetermined time. The dehydrogenization of the asphalt forms hydrogen sulfide, and the oxidation creates sulfur dioxide. Steam, used to blanket the top of the tower to entrain the various contaminants, is then passed through a scrubber to condense the hydrocarbons.

Another process used to produce asphalt is solvent deasphalting. In this extraction process, which uses propane (or hexane) as a solvent, heavy oil fractions are separated to produce heavy lubricating oil, catalytic cracking feedstock, and asphalt. Feedstock and liquid propane are pumped to an extraction tower at precisely controlled mixtures, temperatures (150–250°F), and pressures of 350–600 psi. Separation occurs in a rotating disc contactor, based on differences in solubility. The products are then evaporated and steam stripped to recover the propane, which is recycled. Deasphalting also removes some sulfur and nitrogen compounds, metals, carbon residues, and paraffins from the feedstock.

The potential for a fire exists if a product leak or release contacts a source of ignition such as the process heater. Condensed steam from the various asphalt and deasphalting processes will contain trace amounts of hydrocarbons. Any disruption of the vacuum can result in the entry of atmospheric air and subsequent fire. In addition, raising the temperature of the vacuum tower bottom to improve efficiency can generate methane by thermal cracking. This can create vapors in asphalt storage tanks that are not detectable by flash testing but are high enough to be flammable.

Deasphalting requires exact temperature and pressure control. In addition, moisture, excess solvent, or a drop in operating temperature may cause foaming, which affects the product temperature control and may create an upset. Because these are closed processes, exposures are expected to be minimal during normal operations. Should a spill or release occur, there is a potential for exposure to residuals and asphalt. Air blowing can create some polynuclear aromatics. Condensed steam from the air-blowing asphalt process may also contain contaminants. The potential for exposure to hydrogen sulfide and sulfur dioxide exists in the production of asphalt.

1.5.22 Blending

Blending is the physical mixture of a number of different liquid hydrocarbons to produce a finished product with certain desired characteristics. Products can be blended in-line through a manifold system, or batch blended in tanks and vessels. In-line blending of gasoline, distillates, jet fuel, and kerosene is accomplished by injecting proportionate amounts of each component into the main stream, where turbulence promotes thorough mixing. Additives including octane enhancers, metal deactivators, antioxidants, antiknock agents, gum and rust inhibitors, and detergents are added during and/or after blending to provide specific properties not inherent in hydrocarbons. Ignition sources in the area need to be controlled in the event of a leak or release.

1.5.23 Lubricant, wax, and grease manufacturing

Lubricating oils and waxes are refined from the residual fractions of atmospheric and vacuum distillation. The primary objective of the various lubricating oil refinery processes is to remove asphalts, sulfonated aromatics, and paraffinic and isoparaffinic waxes from residual fractions. Reduced crude from the vacuum unit is deasphalted and combined with straight-run lubricating oil feedstock, preheated, and solvent-extracted (usually with phenol or furfural) to produce raffinate.

In wax manufacturing the raffinate from the extraction unit contains a considerable amount of wax that must be removed by solvent extraction and crystallization. The raffinate is mixed with a solvent (propane) and pre-cooled in heat exchangers. The crystallization temperature is attained by the evaporation of propane in the chiller and filter feed tanks. The wax is continuously removed by filters and cold solvent-washed to recover retained oil. The solvent is recovered from the oil by flashing and steam stripping. The wax is then heated with hot solvent, chilled, filtered, and given a final wash to remove all oil.

In the lubricating oil process, the dewaxed raffinate is blended with other distillate fractions and further treated for viscosity index, color, stability, carbon residue, sulfur, additive response, and oxidation stability in extremely selective extraction processes using solvents (furfural, phenol, etc.). In a typical phenol unit, the raffinate is mixed with phenol in the treating section at temperatures below 400°F. Phenol is then separated from the treated oil and recycled. The treated lube-oil base stocks are then mixed and/or compounded with additives to meet the required physical and chemical characteristics of motor oils, industrial lubricants, and metal working oils.

Grease is made by blending metallic soaps (salts of long-chained fatty acids) and additives into a lubricating oil medium at temperatures of 400–600°F. Grease may be either batch-produced or continuously compounded. The characteristics of the grease depend to a great extent on the metallic element (calcium, sodium, aluminum, lithium, etc.) in the soap and the additives used.

The potential for fire exists if a product or vapor leak or release in the lube-blending and wax-processing areas reaches a source of ignition. Storage of finished products, both bulk and packaged, should be in accordance with recognized practices. While the potential for fire is reduced in lube oil blending, care must be taken when making metalworking oils and compounding greases due to the use of higher blending and compounding temperatures and lower flash-point products. Control of treater temperature is important as phenol can cause corrosion above 400°F. Batch and in-line blending operations require strict controls to maintain desired product quality. Spills should be cleaned and leaks repaired to avoid slips and falls. Additives in drums and bags need to be handled properly to avoid strain. Wax can clog sewer or oil drainage systems and interfere with wastewater treatment. When blending, sampling, and compounding, personal protection from steam, dusts, mists, vapors, metallic salts, and other additives is appropriate. Skin contact with any formulated grease or lubricant should be avoided.

1.5.24 *Heat exchangers, coolers, and process heaters*

Process heaters and heat exchangers preheat feedstock in distillation towers and in refinery processes to reaction temperatures. Heat exchangers use either steam or hot hydrocarbon transferred from some other section of the process for heat input. The heaters are usually designed for specific process operations, and most are of cylindrical vertical or box-type designs. The major portion of heat provided to process units comes from fired heaters fueled by refinery or natural gas, distillate, and residual oils. Fired heaters are found on crude and reformer preheaters, coker heaters, and large-column reboilers.

Heat also may be removed from some processes by air and water exchangers, fin fans, gas and liquid coolers, and overhead condensers, or by transferring heat to other systems. The basic mechanical vapor-compression refrigeration system, which may serve one or more process units, includes an evaporator, compressor, condenser, controls, and piping. Common coolants are water, alcohol/water mixtures, or various glycol solutions.

A means of providing adequate draft or steam purging is required to reduce the chance of explosions when lighting fires in heater furnaces. Specific startup and emergency procedures are required for each type of unit. If fire impinges on fin fans, failure could occur due to overheating. If flammable product escapes from a heat exchanger or cooler due to a leak, fire could occur. Care must be taken to ensure that all pressure is removed from heater tubes before removing header or fitting plugs. Consideration should be given to providing for pressure relief in heat-exchanger piping systems in the event they are blocked off while full of liquid. If controls fail, variations of temperature and pressure could occur on either side of the heat exchanger. If heat exchanger tubes fail and process pressure is greater than heater pressure, product could enter the heater with downstream consequences. If the process pressure is less than heater pressure, the heater stream could enter into the process fluid. If loss of circulation occurs in liquid or gas coolers, increased product temperature could affect downstream operations and require pressure relief. Because these are closed systems, exposures under normal operating conditions are expected to be minimal. Depending on the fuel, process operation, and unit design, there is a potential for exposure to hydrogen sulfide, carbon monoxide, hydrocarbons, steam boiler feed-water sludge, and water-treatment chemicals. Skin contact should be avoided with boiler blowdown, which may contain phenolic compounds.

1.5.25 Steam generation

Steam is generated in main generation plants, and/or at various process units using heat from flue gas or other sources. Heaters (furnaces) include burners and a combustion air system, the boiler enclosure in which heat transfer takes place, a draft or pressure system to remove flue gas from the furnace, soot blowers, and compressed-air systems that seal openings to prevent the escape of flue gas. Boilers consist of a number of tubes that carry the water–steam mixture through the furnace for maximum heat transfer. These tubes run between steam-distribution drums at the top of the boiler and water-collecting drums at the bottom of the boiler. Steam flows from the steam drum to the superheater before entering the steam-distribution system.

Heaters use any one or a combination of fuels including refinery gas, natural gas, fuel oil, and powdered coal. Refinery off-gas is collected from process units and combined with natural gas and LPG in a fuel-gas balance drum. The balance drum provides constant system pressure, fairly stable Btu-content fuel, and automatic separation of suspended liquids in gas vapors, and it prevents carryover of large slugs of condensate into the distribution system. Fuel oil is typically a mix of refinery crude oil with straight-run and cracked residues and other products. The fuel-oil system delivers fuel to process-unit heaters and steam generators at required temperatures and pressures. The fuel oil is heated to pumping temperature, sucked through a coarse suction strainer, pumped to a temperature-control heater, and then pumped through a fine-mesh strainer before being burned.

The distribution system consists of valves, fittings, piping, and connections suitable for the pressure of the steam transported. Steam leaves the boilers at the highest pressure required by the process units or electrical generation. The steam pressure is then reduced in turbines that drive process pumps and compressors. Most steam used in the refinery is condensed to water in various types of heat exchangers. The condensate is reused as boiler feedwater or discharged to wastewater treatment. When refinery steam is also used to drive steam turbine generators to produce electricity, the steam must be produced at much higher pressure than required for process steam. Steam typically is generated by heaters (furnaces) and boilers combined in one unit.

Feedwater supply is an important part of steam generation. There must always be as many pounds of water entering the system as there are pounds of steam leaving it. Water used in steam generation must be free of contaminants including minerals and dissolved impurities that can damage the system or affect its operation. Suspended materials such as silt, sewage, and oil, which form scale and sludge, must be coagulated or filtered out of the water.

Dissolved gases, particularly carbon dioxide and oxygen, cause boiler corrosion and are removed by deaeration and treatment. Dissolved minerals including metallic salts, calcium, carbonates, etc., that cause scale, corrosion, and turbine blade deposits, are treated with lime or soda ash to precipitate them from the water. Recirculated cooling water must also be treated for hydrocarbons and other contaminants. Depending on the characteristics of raw boiler feedwater, some or all of the following stages of treatment will be applicable: clarification; sedimentation; filtration; ion exchange; deaeration; and internal treatment.

The most potentially hazardous operation in steam generation is heater startup. A flammable mixture of gas and air can build up as a result of loss of flame at one or more burners during light-off. Each type of unit requires specific startup and emergency procedures including purging before light-off and in the event of misfire or loss of burner flame. If feedwater runs low and boilers are dry, the tubes will overheat and fail. Conversely, excess water will be carried over into the steam distribution system and damage the turbines. Feedwater must be free of contaminants that could affect operations. Boilers should have continuous or intermittent blowdown systems to remove water from steam drums and limit buildup of scale on turbine blades and superheater tubes. Care must be taken not to overheat the superheater during startup and shutdown. Alternate fuel sources should be provided in the event of loss of gas due to refinery unit shutdown or emergency. Knockout pots provided at process units remove liquids from fuel gas before burning.

1.5.26 *Pressure-relief and flare systems*

Pressure-relief systems control vapors and liquids that are released by pressure-relieving devices and blowdowns. Pressure relief is an automatic, planned release when operating pressure reaches a predetermined level. Blowdown normally refers to the intentional release of material, such as blowdowns from process unit startups, furnace blowdowns, shutdowns, and emergencies. Vapor depressuring is the rapid removal of vapors from pressure vessels in case of fire. This may be accomplished by the use of a rupture disc, usually set at a higher pressure than the relief valve.

Safety relief valves, used for air, steam, and gas as well as for vapor and liquid, allow the valve to open in proportion to the increase in pressure over the normal operating pressure. Safety valves designed primarily to release high volumes of steam usually pop open to full capacity. The overpressure needed to open liquid-relief valves where large-volume discharge is not required increases as the valve lifts due to increased spring resistance. Pilot-operated safety relief valves, with up to six times the capacity of normal relief valves, are used where tighter sealing and larger volume discharges are required. Nonvolatile liquids are usually pumped to oil–water separation and recovery systems, and volatile liquids are sent to units operating at a lower pressure.

A typical closed pressure release and flare system includes relief valves and lines from process units for collection of discharges, knockout drums to separate vapors and liquids, seals, and/or purge gas for flashback protection, and a flare and igniter system that combusts vapors when discharging directly to the atmosphere is not permitted. Steam may be injected into the flare tip to reduce visible smoke.

Vapors and gases must not discharge where sources of ignition could be present. Liquids should not be discharged directly to a vapor disposal system. Flare knockout drums and flares need to be large enough to handle emergency blowdowns. Drums should be provided with relief in the event of overpressure. Pressure relief valves must be provided where the potential exists for overpressure in refinery processes due to the following causes:

- loss of cooling water, which may greatly reduce pressure in condensers and increase the pressure in the process unit;
- loss of reflux volume, which may cause a pressure drop in condensers and a pressure rise in distillation towers because the quantity of reflux affects the volume of vapors leaving the distillation tower;
- rapid vaporization and pressure increase from injection of a lower-boiling-point liquid including water into a process vessel operating at higher temperatures;
- expansion of vapor and resultant overpressure due to overheated process steam, malfunctioning heaters, or fire;
- failure of automatic controls, closed outlets, heat exchanger failure, etc;
- internal explosion, chemical reaction, thermal expansion, or accumulated gases.

Maintenance is important because valves are required to function properly. The most common operating problems are listed below.

- failure to open at set pressure, because of plugging of the valve inlet or outlet, or because corrosion prevents proper operation of the disc holder and guides;
- failure to reseat after popping open due to fouling, corrosion, or deposits on the seat or moving parts, or because solids in the gas stream have cut the valve disc;
- chattering and premature opening, because operating pressure is too close to the set point.

In general, petroleum refineries are faced with the problem of safe disposal of volatile liquids and gases resulting from scheduled shutdowns and sudden or unexpected upsets in process units. Emergencies that can cause the sudden venting of excessive amounts of gases and vapors include fires, compressor failures, overpressures in process vessels, line breaks, leaks, and power failures. As noted, uncontrolled releases of large volumes of gases also constitute a serious safety hazard to personnel and equipment. A system for disposal of emergency and waste refinery gases consists of a manifolded pressure-relieving or blowdown system, and a blowdown recovery system or a system of flares for the combustion of the excess gases, or both. Many older refineries, however, do not operate blowdown recovery systems. In addition to disposing of emergency and excess gas flows, these systems are used in the evacuation of units during shutdowns and turnarounds. Normally a unit is shut down by depressuring into a fuel gas or vapor recovery system, with further depressuring to essentially atmospheric pressure by venting to a low-pressure flare system. Thus, overall emissions of refinery hydrocarbons are substantially reduced.

Blowdown systems are used to ensure the safety of personnel and protect equipment in the event of emergencies such as process upset, equipment failure, and fire. A properly designed pressure-relief system permits substantial reduction of hydrocarbon emissions to the atmosphere. The equipment in a refinery can operate at pressures ranging from less than atmospheric to 1000 psig and higher. This equipment must be designed to permit safe disposal of excess gases and liquids in case operational difficulties or fires occur. These materials are usually removed from the process area by automatic safety and relief valves, as well as by manually controlled valves, manifolded to a header that conducts the material away from the unit involved.

One of the preferred methods of disposing of the waste gases that cannot be recovered in a blowdown recovery system is by burning in a smokeless flare. Liquid blowdowns are usually sent to appropriately designed holding vessels and reclaimed. A blowdown or pressure-relieving system consists of relief valves, safety valves, manual bypass valves, blowdown headers, knockout vessels, and holding tanks. A blowdown recovery system also includes compressors and vapor surge vessels such as gas holders or vapor spheres.

Flares are usually considered as part of the blowdown system in a modern refinery. The pressure-relieving system can be used for liquids or vapors or both. For reasons of economy and safety, vessels and equipment discharging to blowdown systems are usually segregated according to their operating pressure. In other words, there is a high-pressure blowdown system for equipment working, for example, above 100 psig, and low-pressure systems for those vessels with working pressures below 100 psig. Butane and propane are usually discharged to a separate blowdown drum, which is operated above atmospheric pressure to increase recovery of liquids. Usually a direct-contact type of condenser is used to permit recovery of as much hydrocarbon liquid as possible from the blowdown vapors. The non-condensables are burned in a flare system.

A typical pressure-relieving system for flaring operations is used not only as a safety measure but also as a means of reducing the emission of hydrocarbons to the atmosphere. A typical installation includes four separate collecting systems as follows: (1) a low-pressure blowdown system for vapors from equipment with working pressures below 100 psig; (2) a high-pressure blowdown system for vapors from equipment with working pressures above 100 psig; (3) a liquid blowdown system for liquids at all pressures; and

(4) a light-ends blowdown for butanes and lighter hydrocarbon blowdown products. The liquid portion of light hydrocarbon products released through the light-ends blowdown system is recovered in a drum near the flare. A backpressure of 50 psig is maintained on the drum, which minimizes the amount of vapor that vents through a backpressure regulator to the high-pressure blowdown line. The high-pressure, low-pressure, and liquid-blowdown systems all discharge into the main blowdown vessel. Any entrained liquid is dropped out and pumped to a storage tank for recovery. Off-gas from this blowdown drum flows to a vertical vessel with baffle trays in which the gases are contacted directly with water, which condenses some of the hydrocarbons and permits their recovery. The overhead vapors from the sump tank flow to the flare system manifold for disposal by burning in a smokeless flare system.

The design of a pressure-relief system is one of the most important problems in the planning of a refinery or petrochemical plant. The safety of personnel and equipment depends on the proper design and functioning of this type of system. The consequences of poor design can be disastrous. A pressure-relief system can consist of one relief valve, safety valve, or rupture disc, or of several relief devices manifolded to a common header. Usually the systems are segregated according to the type of material handled, i.e. liquid or vapor, as well as to the operating pressures involved.

The several factors that must be considered in designing a pressure-relief system are: (1) the governing code, such as that of the American Society of Mechanical Engineers (ASME); (2) characteristics of the pressure-relief devices; (3) the design pressure of the equipment protected by the pressure-relief devices; (4) line sizes and lengths; and (5) physical properties of the material to be relieved to the system. In discussing pressure-relief systems, the following terms are commonly used:

- **Relief valve.** A relief valve is an automatic pressure-relieving device actuated by the static pressure upstream of the valve. It opens further with increase of pressure over the set pressure. It is used primarily for liquid service.
- **Safety valve.** A safety valve is an automatic relieving device actuated by the static pressure upstream of the valve and characterized by full opening or pop action upon opening. It is used for gas or vapor service.
- **Rupture disc.** A rupture disc consists of a thin metal diaphragm held between flanges.
- **Maximum allowable working pressure.** The maximum allowable working pressure (i.e. design pressure), as defined in the construction codes for unfired pressure vessels, depends upon the type of material, its thickness, and the service condition set as the basis for design. The vessel may not be operated above this pressure or its equivalent at any metal temperature higher than that used in its design; consequently, for that metal temperature, it is the highest pressure at which the primary safety or relief valve may be set to open.
- **Operating pressure.** The operating pressure of a vessel is the pressure, in psig, to which the vessel is usually subjected in service. A processing vessel is usually designed

to a maximum allowable working pressure, in psig, that will provide a suitable margin above the operating pressure in order to prevent any undesirable operation of the relief valves. It is suggested that this margin be approximately 10% higher, or 25 psi, whichever is greater.

- **Set pressure.** The set pressure, in psig, is the inlet pressure at which the safety or relief valve is adjusted to open.
- **Accumulation.** Accumulation is the pressure increase over the maximum allowable working pressure of the vessel during discharge to the safety or relief valve expressed as a percentage of that pressure or pounds per square inch.
- **Overpressure.** Overpressure is the pressure increase over the set pressure of the primary relieving device. It is the same as accumulation when the relieving device is set at the maximum allowable working pressure of the vessel. When the set pressure of the first safety or relief valve to open is less than the maximum allowable working pressure of the vessel, the overpressure may be greater than 10% of the set pressure of the first safety or relief valve.
- **Blowdown.** Blowdown is the difference between the set pressure and the reseating pressure of a safety or relief valve, expressed as a percentage of a set pressure or pounds per square inch.
- **Lift.** Lift is the rise of the disc in a safety or relief valve.
- **Backpressure.** Backpressure is the pressure developed on the discharge side of the safety valves. Superimposed backpressure is the pressure in the discharge header before the safety valve opens (discharged from other valves).
- **Built-up pressure.** Built-up backpressure is the pressure in the discharge header after the safety valve opens.

Safety valves

Nozzle-type safety valves are available in the conventional or balanced-bellows configurations. Backpressure in the piping downstream of the standard-type valve affects its set pressure, but theoretically, this backpressure does not affect the set pressure of the balanced-type valve. Owing, however, to imperfections in manufacture and limitations of practical design, the balanced valves available vary in relieving pressure when the backpressure reaches approximately 40% of the set pressure. The actual accumulation depends upon the manufacturer.

Until the advent of balanced valves, the general practice in the industry was to select safety valves that start relieving at the design pressure of the vessel and reach full capacity at 3–10% above the design pressure. This overpressure was defined as accumulation. With the balanced safety valves, the allowable accumulation can be retained with smaller pipe size. Each safety valve installation is an individual problem. The required capacity of the valve depends upon the condition producing the overpressure.

Rupture discs

A rupture disc is an emergency relief device consisting of a thin metal diaphragm carefully designed to rupture at a predetermined pressure. The obvious difference between a relief or safety valve and a rupture disc is that the valve reseats and the disc does not. Rupture discs may be installed in parallel or series with a relief valve. To prevent an incorrect pressure differential from existing, the space between the disc and the valve must be maintained at atmospheric pressure. A rupture disc is usually designed to relieve at 1.5 times the maximum allowable working pressure of the vessel. In determining the size of a disc, three important effects that must be evaluated are low rupture pressure, elevated temperatures, and corrosion. Manufacturers can supply discs that are guaranteed to burst at ±5% of their rated pressures. The corrosive effects of a system determine the type of material used in a disc. Even a slight amount of corrosion can drastically shorten disc life. Discs are available with plastic linings, or they can be made from pure carbon materials.

The discharge piping for relief and safety valves and rupture discs should have a minimum of fittings and bends. There should be minimum loading on the valve, and piping should be used with adequate supports and expansion joints. Suitable drains should be used to prevent liquid accumulation in the piping and valves.

Flares

Smoke is the result of incomplete combustion. Smokeless combustion can be achieved by: (1) adequate heat values to obtain the minimum theoretical combustion temperatures; (2) adequate combustion air; and (3) adequate mixing of the air and fuel. An insufficient supply of air results in a smoky flame. Combustion begins around the periphery of the gas stream where the air and fuel mix, and within this flame envelope the supply of air is limited. Hydrocarbon side reactions occur with the production of smoke. In this reducing atmosphere, hydrocarbons crack to elemental hydrogen and carbon, or polymerize to form hydrocarbons. Since the carbon particles are difficult to burn, large volumes of carbon particles appear as smoke upon cooling.

Side reactions become more pronounced as molecular weight and unsaturation of the fuel gas increase. Olefins, diolefins, and aromatics characteristically burn with smoky, sooty flames as compared with paraffins and naphthenes. A smokeless flame can be obtained when an adequate amount of combustion air is mixed sufficiently with the fuel so that it burns completely and rapidly before any side reactions can take place. Combustion of hydrocarbons in the steam-inspirated-type elevated flare appears to be complete. The air pollution problem associated with the uncontrolled disposal of waste gases is the venting of large volumes of hydrocarbons and other odorous gases and aerosols. The preferred control method for excess gases and vapors is to recover them in a blowdown recovery system and, failing that, to incinerate them in an elevated-type flare. Such flares introduce the possibility of smoke and other objectionable gases such as carbon monoxide, sulfur dioxide, and nitrogen oxides. Flares have been further developed to ensure that this combustion is smokeless and in some cases nonluminous. Luminosity does attract attention to the refinery operation and in certain cases can cause bad public relations. There is also the consideration of military security in which nonluminous emergency gas flares would be desirable. It is important to note that the hydrocarbon and carbon monoxide emissions from a flare can be much greater than those from a properly operated refinery boiler or furnace. Other combustion contaminants from a flare include nitrogen oxides. The importance of these compounds to the total air pollution problem depends upon the particular conditions in a particular locality.

Other air contaminants that can be emitted from flares depend upon the composition of the gases burned. The most commonly detected emission is sulfur dioxide, resulting from the combustion of various sulfur compounds (usually hydrogen sulfide) in the flared gas. Toxicity, combined with low odor threshold, make venting of hydrogen sulfide to a flare an unsuitable and sometimes dangerous method of disposal. In addition, burning relatively small amounts of hydrogen sulfide can create enough sulfur dioxide to cause crop damage or local nuisance. Materials that tend to cause health hazards or nuisances should not be disposed of in flares. Compounds such as mercaptans or chlorinated hydrocarbons require special combustion devices with chemical treatment of the gas or its products of combustion.

The ideal refinery flare is a simple device for safe and inconspicuous disposal of waste gases by combustion. Hence, the ideal flare is a combustion device that burns waste gases completely and smokelessly. There are, in general, three types of flare: elevated flares, ground-level flares, and burning pits. The burning pits are reserved for extremely large gas flows caused by catastrophic emergencies in which the capacity of the primary smokeless flares is exceeded. Ordinarily, the main gas header to the flare system has a water seal bypass to a burning pit. Excessive pressure in the header blows the water seal and permits the vapors and gases to vent a burning pit where combustion occurs. This is rarely practiced today, except for parts of South America and Eastern Europe.

The essential parts of a flare are the burner, stack, seal, liquid trap, controls, pilot burner, and ignition system. In some cases, vented gases flow through chemical solutions to receive treatment before combustion. As an example, gases vented from an isomerization unit that may contain small amounts of hydrochloric acid are scrubbed with caustic before being vented to the flare.

Elevated flares are the most commonly used system. Smokeless combustion can be obtained in an elevated flare by the injection of an inert gas to the combustion zone to provide turbulence and inspirate air. A mechanical air-mixing system would be ideal but is not economical in view of the large volume of gases typically handled. The most commonly encountered air-inspirating material for an elevated flare is steam. Three main types of steam-injected elevated flares are in use. These types vary in the manner in which the steam is injected into the combustion zone. In the first type, there is a commercially available multiple nozzle that consists of an alloy steel tip mounted on the top of an elevated stack. Steam injection is accomplished by several small jets placed concentrically around the flare tip. These jets are installed at an angle, causing the steam to discharge in a converging pattern immediately above the flare tip. A second type of elevated flare has a flare tip with no obstruction to flow, i.e. the flare tip has the same diameter as the stack. The steam is injected by a single nozzle located concentrically within the burner tip. In this type of flare, the steam is premixed with the gas before ignition and discharge.

A third type of elevated flare is equipped with a flare tip constructed to cause the gases to flow through several tangential openings to promote turbulence. A steam ring at the top of the stack has numerous equally spaced holes about {1/8} inch in diameter for discharging steam into the gas stream.

The injection of steam in this latter flare may be automatically or manually controlled. In most cases, the steam is proportioned automatically to the rate of gas flow; however, in some installations, the steam is automatically supplied at maximum rates, and manual throttling of a steam valve is required for adjusting the steam flow to the particular gas flow rate. There are many variations of instrumentation among various flares, some designs being more desirable than others.

For economic reasons all designs attempt to proportion steam flow to the gas flow rate. Steam injection is generally believed to result in the following benefits: (1) energy available at relatively low cost can be used to inspirate air and provide turbulence within the flame; (2) steam reacts with the fuel to form oxygenated compounds that burn readily at relatively low temperatures; (3) water–gas reactions also occur with this same end-result; and (4) steam reduces the partial pressure of the fuel and retards polymerization. Inert gases such as nitrogen have also been found effective for this purpose; however, the expense of providing a diluent such as this is prohibitive.

There are four principal types of ground-level flare: horizontal venturi, water injection, multijet, and vertical venturi. In a typical horizontal, venturi-type ground flare system, the refinery flare header discharges to a knockout drum where any entrained liquid is separated and pumped to storage. The gas flows to the burner header, which is connected to three separate banks of standard gas burners through automatic valves of the snap-action type that open at predetermined pressures. If any or all of the pressure valves fail, a bypass line with a liquid seal is provided (with no valves in the circuit), which discharges to the largest bank of burners. The automatic-valve operation schedule is determined by the quantity of gas most likely to be relieved to the system.

The allowable backpressure in the refinery flare header determines the minimum pressure for the control valve and the No. 1 burner bank. On the assumption that the first valve was set at 3 psig, then the second valve for the No. 2 burner bank would be set for some higher pressure, say 5 psig. The quantity of gas most likely to be released then determines the size and the number of burners for this section. Again, the third most likely quantity of gas determines the pressure setting and the size of the third control valve. Together, the burner capacity should equal the maximum expected flow rate. The valve-operating schedule for the system is set up as follows. (1) When the relief header pressure reaches 3 psig, the first control valve opens and the four small venturi burners go into operation. The controller setting keeps the valve open until the pressure decreases to about 1.5 psig. (2) When the header pressure reaches 5 psig, the second valve opens and remains open until the pressure drops to about 3 psig. (3) When the pressure reaches 6 psig, the third valve opens and remains open until the pressure decreases to 4 psig. (4) At about 7 psig, the gas blows the liquid seal.

Another common type of ground flare used in petroleum refineries has a water spray to inspirate air and provide water vapor for the smokeless combustion of gases. This flare requires an adequate supply of water and a reasonable amount of open space. The structure of the flare consists of three concentric stacks. The combustion chamber contains the burner, the pilot burner, the end of the igniter tube, and the water spray distributor ring. The primary purpose of the intermediate stack is to combine the water spray with burning gases. The outer stack confines the flame and directs it upward. Water sprays in elevated flares are not too practical for several reasons. Difficulty is experienced in keeping the water spray in the flame zone, and the scale formed in the waterline tends to plug the nozzles. Water main pressure dictates the height to which water can be injected without the use of a booster pump. For a 100- to 250-foot stack, a booster pump would undoubtedly be required. Rain created by the spray from the flare stack is objectionable from the standpoint of corrosion of nearby structures and other equipment. Water is not as effective as steam for controlling smoke with high gas flow rates, unsaturated materials, or wet gases. The water spray flare is economical when venting rates are not too high and slight smoking can be tolerated.

The multijet-type ground flare is designed to burn excess hydrocarbons without smoke, noise, or visible flame. These generally tend to be less expensive than the steam-injected type, on the assumption that new steam facilities must be installed to serve a steam-injected flare unit. Where the steam can be diverted from noncritical operations such as tank heating, the cost of the multijet flare and the steam inspirating elevated flare may be similar. These flares use two sets of burners; the smaller group handles normal gas leakage and small gas releases, while both burner groups are used at higher flaring rates. This sequential operation is controlled by two water-sealed drums set to release at different pressures. In extreme emergencies, the multijet burners are bypassed by means of a water seal that directs the gases to the center of the stack. This seal blows at flaring rates higher than the design capacity of the flare. At such an excessive rate, the combustion is both luminous and smoky, but the unit is usually sized so that an overcapacity flow would be a rare occurrence. The overcapacity line may also be designed to discharge through a water seal to a nearby elevated flare rather than to the center of a multijet stack. Similar staging could be accomplished with automatic valves or back-pressure regulators; however, in this case, the water seal drums are used because of reliability and ease of maintenance. The staging system is balanced by adjusting the hand control butterfly valve leading to the first-stage drum. After its initial setting, this valve is locked into position. The vertical, venturi-type ground flare is a design based upon the use of commercial-type venturi burners. This type of flare has been used to handle vapors from gas-blanketed tanks, and vapors displaced from the depressuring of butane and propane tank trucks. Since the commercial venturi burner requires a certain minimum pressure to operate efficiently, a gas blower must be provided. A compressor takes vapors from tankage and discharges them through a water seal tank and a flame arrestor to the flare. This type of arrangement can readily be modified to handle different volumes of vapors by the installation of the necessary number of burners. This type of flare is suitable for relatively small flows of gas of a constant rate. Its main application is in situations where other means of disposing of gases and vapors are not available.

A flare installation that does not inspirate an adequate amount of air or does not mix the air and hydrocarbons properly emits dense, black clouds of smoke that obscure the flame. The injection of steam into the zone of combustion causes a gradual decrease in the amount of smoke, and the flame becomes more visible.

When trailing smoke has been eliminated, the flame is very luminous and orange with a few wisps of black smoke around the periphery. The minimum amount of steam required produces a yellowish orange, luminous flame with no smoke. Increasing the amount of steam injection further decreases the luminosity of the flame. As the steam rate increases, the flame becomes colorless and finally invisible during the day. At night this flame appears blue. An injection of an excessive amount of steam causes the flame to disappear completely and be replaced with a steam plume. An excessive amount of steam may extinguish the burning gases and permit unburned hydrocarbons to discharge to the atmosphere. When the flame is out, there is a change in the sound of the flare because a steam hiss replaces the roar of combustion.

The commercially available pilot burners are usually not extinguished by excessive amounts of steam, and the flame reappears as the steam injection rate is reduced. As the use of automatic instrumentation becomes more prevalent in flare installations, the use of excessive amounts of steam and the emission of unburned hydrocarbons decrease and greater steam economies can be achieved. In evaluating flare installations, controlling the volume of steam is important. Too little steam results in black smoke, which obviously is objectionable. Conversely, excessive use of steam produces a white steam plume and an invisible emission of unburned hydrocarbons. A condition such as this can also be a serious air pollution problem.

The venturi-type ground flare, as previously discussed, consists of burners, pilots, igniters, and control valves. The total pressure drop permitted in a given installation depends upon the characteristics of the particular blow-down system. In general, the allowable pressure drop through the relief valve headers, liquid traps, burners, and so forth must not exceed one-half the internal unit's relieving pressure. The burner cut-in schedule is based upon a knowledge of the source, frequency, and quantity of the release gases. Pressure downstream of the control valves must be adequate to provide stable burner operation.

Flare installations designed for relatively small gas flows can use clusters of commercially available venturi burners. For large gas releases, special venturi burners must be constructed. The venturi (air-inspirating) burners are installed in clusters with a small venturi-type pilot burner in the center. This burner should be connected to an independent gas source. The burners may be mounted vertically or horizontally.

The burners should fire through a refractory wall to provide protection for personnel and equipment. Controls can be installed to give remote indication of the pilot burner's operation. For large-capacity venturi burners, field tests are necessary to obtain the proper throat-to-orifice ratio and the minimum pressure for stable burner operation.

1.5.27 Wastewater treatment

Wastewater treatment is used for process, runoff, and sewerage water prior to discharge or recycling. Wastewater typically contains hydrocarbons, dissolved materials, suspended solids, phenols, ammonia, sulfides, and other compounds. Wastewater includes condensed steam, stripping water, spent caustic solutions, cooling tower and boiler blowdown, wash water, alkaline and acid waste neutralization water, and other process-associated water.

Pretreatment is the separation of hydrocarbons and solids from wastewater. API separators, interceptor plates, and settling ponds remove suspended hydrocarbons, oily sludge, and solids by gravity separation, skimming, and filtration. Some oil-in-water emulsions must be heated to assist in separating the oil and water. Gravity separation depends on the specific gravity differences between water and immiscible oil globules and allows free oil to be skimmed off the surface of the wastewater. Acidic wastewater is neutralized using ammonia, lime, or soda ash. Alkaline wastewater is treated with sulfuric acid, hydrochloric acid, carbon dioxide-rich flue gas, or sulfur.

After pretreatment, suspended solids are removed by sedimentation or air flotation. Wastewater with low levels of solids may be screened or filtered. Flocculation agents are sometimes added to help separation. Secondary treatment processes biologically degrade and oxidize soluble organic matter by the use of activated sludge, unaerated or aerated lagoons, trickling filter methods, or anaerobic treatments. Materials with high adsorption characteristics are used in fixed-bed filters or added to the wastewater to form a slurry that is removed by sedimentation or filtration. Additional treatment methods are used to remove oils and chemicals from wastewater. Stripping is used on wastewater containing sulfides and/or ammonia, and solvent extraction is used to remove phenols.

Tertiary treatments remove specific pollutants to meet regulatory discharge requirements. These treatments include chlorination, ozonation, ion exchange, reverse osmosis, and activated carbon adsorption. Compressed oxygen is diffused into wastewater streams to oxidize certain chemicals or to satisfy regulatory oxygen-content requirements. Wastewater that is to be recycled may require cooling to remove heat and/or oxidation by spraying or air stripping to remove any remaining phenols, nitrates, and ammonia.

The potential for fire exists if vapors from wastewater containing hydrocarbons reach a source of ignition during treatment.

1.5.28 Cooling towers

Cooling towers remove heat from process water by evaporation and latent heat transfer between hot water and air. The two types of towers are crossflow and counterflow. Crossflow towers introduce the airflow at right angles to the water flow throughout the structure. In counterflow cooling towers, hot process water is pumped to the uppermost plenum and allowed to fall through the tower. Numerous slats or spray nozzles located throughout the length of the tower disperse the water and help in cooling. Air enters at the tower bottom and flows upward against the water. When the fans or blowers are at the air inlet, the air is considered to be forced draft. Induced draft is when the fans are at the air outlet.

Recirculated cooling water must be treated to remove impurities and dissolved hydrocarbons. Because the water is saturated with oxygen from being cooled with air, the chances for corrosion are increased. One means of corrosion prevention is the addition of a material to the cooling water that forms a protective film on pipes and other metal surfaces.

When cooling water is contaminated by hydrocarbons, flammable vapors can be evaporated into the discharge air. If a source of ignition is present, or if lightning occurs, a fire may start. A potential fire hazard also exists where there are relatively dry areas in induced-draft cooling towers of combustible construction. Loss of power to cooling tower fans or water pumps could have serious consequences in the operation of the refinery. Impurities in cooling water can corrode and foul pipes and heat exchangers, scale from dissolved salts can deposit on pipes, and wooden cooling towers can be damaged by microorganisms. Cooling-tower water can be contaminated by process materials and by-products including sulfur dioxide, hydrogen sulfide, and carbon dioxide, with resultant exposures. Safe work practices and/or appropriate personal protective equipment may be needed during process sampling, inspection, maintenance, and turnaround activities, and for exposure to hazards such as those related to noise, water-treatment chemicals, and hydrogen sulfide when wastewater is treated in conjunction with cooling towers.

1.5.29 Gas and air compressors, and turbines

Both reciprocating and centrifugal compressors are used throughout the refinery for gas and compressed air. Air compressor systems include compressors, coolers, air receivers, air driers, controls, and distribution piping. Blowers are used to provide air to certain processes. Plant air is provided for the operation of air-powered tools, catalyst regeneration, process heaters, steam–air decoking, sour-water oxidation, gasoline sweetening, asphalt blowing, and other uses. Instrument air is provided for use in pneumatic instruments and controls, air motors and purge connections.

The OSHA recommends that air compressors be located so that the suction does not take in flammable vapors or corrosive gases. There is a potential for fire should a leak occur in gas compressors. Knockout drums are needed to prevent liquid surges from entering gas compressors. If gases are contaminated with solid materials, strainers are needed. Failure of automatic compressor controls will affect processes. If maximum pressure could potentially be greater than compressor or process-equipment design pressure, pressure relief should be provided. Guarding is needed for exposed moving parts on compressors. Compressor buildings should be properly electrically classified, and provisions should be made for proper ventilation. Where plant air is used to back up instrument air, interconnections must be upstream of the instrument air-drying system to prevent contamination of instruments with moisture. Alternate sources of instrument air supply, such as use of nitrogen, may be needed in the event of power outages or compressor failure. Turbines are usually gas- or steam-powered and are typically used to drive pumps, compressors, blowers, and other refinery process equipment. Steam enters turbines at high temperatures and pressures, expands across and drives rotating blades while being directed by fixed blades.

Steam turbines used for exhaust operating under vacuum should have safety relief valves on the discharge side, both for protection and to maintain steam in the event of vacuum failure. Where maximum operating pressure could be greater than design pressure, steam turbines should be provided with relief devices. Consideration should be given to providing governors and overspeed control devices on turbines.

1.5.30 Marine, tank car, and tank truck loading and unloading

Facilities for loading liquid hydrocarbons into tank cars, tank trucks, and marine vessels and barges are usually part of the refinery operations. Product characteristics, distribution needs, shipping requirements, and operating criteria are important when designing loading facilities. Tank trucks and rail tank cars are either top- or bottom-loaded, and vapor-recovery systems may be provided where required. Loading and unloading liquefied petroleum gas (LPG) require special considerations in addition to those for liquid hydrocarbons.

The potential for fire exists where flammable vapors from spills or releases can reach a source of ignition. Where switch loading is permitted, safe practices need to be established and followed. Bonding is used to equalize the electrical charge between the loading rack and the tank truck or tank car. Grounding is used at truck and rail loading facilities to prevent flow of stray currents. Insulating flanges are used on marine dock piping connections to prevent static electricity buildup and discharge. Flame arrestors should be installed in loading rack and marine vapor-recovery lines to prevent flashback. Automatic or manual shutoff systems at supply headers are needed for top and bottom loading in the event of leaks or overfills. Fall protection such as railings are needed for top-loading racks where employees are exposed to falls. Drainage and recovery systems may be provided for storm drainage and to handle spills and leaks. Precautions must be taken at LPG loading facilities not to overload or overpressurize tank cars and trucks. The nature of the health hazards at loading and unloading facilities depends upon the products being loaded and the products previously transported in the tank cars, tank trucks, or marine vessels. Safe work practices and/or appropriate personal protective equipment may be needed to protect against hazardous exposures when loading or unloading, cleaning up spills or leaks, or when gauging, inspecting, sampling, or performing maintenance activities on loading facilities or vapor-recovery systems.

1.5.31 Pumps, piping, and valves

Centrifugal and positive-displacement (i.e. reciprocating) pumps are used to move hydrocarbons, process water, fire water, and wastewater through piping within the refinery. Pumps are driven by electric motors, steam turbines, or internal combustion engines. The pump type, capacity, and construction materials depend on the service for which it is used. Process and utility piping distribute hydrocarbons, steam, water, and other products throughout the facility. Their size and construction depend on the type of service, pressure, temperature, and nature of the products. Vent, drain, and sample connections are provided on piping, as well as provisions for blanking. Different types of valve are used depending on their operating purpose. These include gate valves, bypass valves, globe and ball valves, plug valves, block and bleed valves, and check valves. Valves can be manually or automatically operated. A refinery may have tens of thousands of valves fitting connectors and various components, all of which are sources of fugitive emissions.

The potential for fire exists should hydrocarbon pumps, valves, or lines develop leaks that could allow vapors to reach sources of ignition. Remote sensors, control valves, fire valves, and isolation valves should be used to limit the release of hydrocarbons at pump suction lines in the event of leakage and/or fire. Depending on the product and service, backflow prevention from the discharge line may be needed. The failure of automatic pump controls could cause a deviation in process pressure and temperature. Pumps operated with reduced or no flow can overheat and rupture. Pressure relief in the discharge piping should be provided where pumps can be overpressured. Provision may be made for pipeline expansion, movement, and temperature changes to avoid rupture. Valves and instruments that require servicing or other work should be accessible at grade level or from an operating platform. Operating vent and drain connections should be provided with double-block valves, a block valve and plug, or blind flange for protection against releases.

1.5.32 Tanks

Atmospheric storage tanks and pressure storage tanks are used throughout the refinery for storage of crudes, intermediate hydrocarbons (during the process), and finished products. Tanks are also provided for fire water, process and treatment water, acids, additives, and other chemicals. The type, construction, capacity, and location of tanks depend on their use and materials stored.

The potential for fire exists should hydrocarbon storage tanks be overfilled or develop leaks that allow vapors to escape and reach sources of ignition. Remote sensors, control valves, isolation valves, and fire valves may be provided at tanks for pump-out or closure in the event of a fire in the tank, or in the tank dike or storage area. Tanks may be provided with automatic overflow control and alarm systems, or manual gauging and checking procedures may be established to control overfills.

1.6 Further reading

There are many fine reference materials both in printed form and on the Web that the reader may access for additional information on refining processes and characterizations of waste streams and emissions. We have found the following references to provide some useful information and recommend them as general references for readers.

American Petroleum Institute, 1993. Environmental Design Considerations for Petroleum Refining Crude Processing Units. API Publication No. 311. February.
Gary, J.C., Handwerk, G.E., 1994. Petroleum Refining – Technology and Economics, 3rd edn. Marcel Dekker, New York.

Leeman, J.E., 1988. Hazardous Waste Minimization: Part V. Waste Minimization in the Petroleum Industry. JAPCA 38 (6) June.

Meyers, R.A., 1986. Handbook of Petroleum Refining Processes. McGraw-Hill, New York.

US Petroleum Refining, 1993. Meeting Requirements for Cleaner Fuels and Refineries, Vols. I–VI. National Petroleum Council Committee on Refining, US Department of Refining, US Department of Energy, Washington, DC. August.

References

Agency for Toxic Substances and Diseases Registry, 2007. ToxFAQs for Methyl tert-Butyl Ether (MTBE), 11 September. Retrieved 16 August 2008, from http://www.atsdr.cdc.gov/tfacts91.html.

Aiden, R., 1958. Petrol-vapor Poisoning. British Medical Journal ii, 369–370.

Ainsworth, R., 1960. Petrol-vapor Poisoning. British Medical Journal 1, 1547–1548.

American Petroleum Institute (API), 1948. Toxicological Review: Benzene. Department of Safety, API, New York. September.

Askey, J.M., 1928. Aplastic Anemia Due to Benzol Poisoning. Calif. West. Med. 29, 262–263.

Baker, C. et al. v. Chevron USA, Inc., Civil Action No. 1:05CV227.

Eaton, W.S., et al., 1980. Fugitive Hydrocarbon Emissions from Petroleum Production Operation (2 volumes). API Publication No. 4322. American Petroleum Institute, Washington, DC. March.

Energy Information Administration, 2006. Office of Oil and Gas, January.

Energy Information Administration, 2008. US MTBE Oxygenate Merchant Facilities Production (Thousand Barrels), 23 July. Retrieved 17 August 2008, from http://tonto.eia.doe.gov/dnav/pet/hist/m_epooxt_ypm_nus_1A.htm.

Hinck, J., 2001. Testing the Waters in MTBE Litigation, 1 July. Retrieved 18 August 2008, from http://www.accessmylibrary.com/coms2/summary_0286-10494133_ITM.

Hummel, K., 1990. Technical Memorandum to C.C. Masser (US Environmental Protection Agency) concerning Screening and Bagging of Selected Fugitive Sources at Natural Gas Production and Processing Facilities, June.

Infante, P.F., Schwartz, E., Cahill, R., 1990. Benzene in Petrol: A Continuing Hazard. Lancet 335, 814–815.

Ivanhoe, L.F., Leckie, G.G., 1993. Global Oil, Gas Fields, Sizes Tallied, Analyzed. Oil and Gas Journal, 87–91. 15 February.

Kovarik, W., 2005. Ethyl-leaded Gasoline: How a Classic Occupational Disease Became an International Public Health Disaster. Int. J. Occup. Environ. Health 11 (4) Oct/Dec.

Serne, J. C., Bernstiel, T. J. and Shermaria, M. A., 1991. *An Air Toxics and VOC Emission Factor Development Project for Oil Production Facilities*. Presented at the 1991 Annual Meeting of the Air and Waste Management Association, Vancouver, British Columbia, Canada.

Takamiya, M., Niitsu, H., Saigusa, K., Kanetake, J., Aoki, Y., 2003. A Case of Acute Gasoline Intoxication at the Scene of Washing a Petrol Tank. Leg. Med. (Tokyo) 5, 165–169.

US Environmental Protection Agency (EPA), 1995. Protocol for Equipment Leak Emission Estimates, EPA-453/R-95–017. US Environmental Protection Agency, Office of Air Quality Planning and Standards, Research Triangle Park, NC. November.

US Environmental Protection Agency (EPA), 2008. Methyl Tertiary Butyl Ether Research, 1 May. Retrieved 17 August 2008, from http://www.epa.gov/nrmrl/lrpcd/esm/mtbe_research.htm.

US Geological Survey, 2007. Methyl Tertiary-Butyl Ether (MTBE), 25 April. Retrieved 16 August 2008, from http://ca.water.usgs.gov/mtbe/index.html.

Waxman report, 1999. Oil Refineries Fail to Report Millions of Pounds of Harmful Emissions. Report prepared for Rep. Henry A. Waxman, Minority Staff: Special Investigations Division, Committee on Government Reform, US House of Representatives, November.

2 The Santa Maria oil sumps

2.1 Introduction

This chapter provides a case study. Santa Maria, a coastal city in Santa Barbara County, sits on top of an oil field. Since 1930, the oil field has been operated by many different oil companies and produced 206 million barrels of oil (California Department of Conservation, 2007). In the 1950s, large oil-well sumps were built to collect by-products of drilling, including water, drilling mud, and oil. Each oil well had at least one sump, varying in size from the size of a house to the size of a football field (Doane-Allmon, 2005).

After its peak in oil production in the 1950s, parts of the oilfield began being decommissioned and the city of Santa Maria began to grow on top of it. Over the next couple of decades, 1707 oil wells were abandoned. As wells were decommissioned, the responsible oil company removed the oil and covered the sumps with 1–4 feet of clean soil (Santa Barbara County Fire Department, 2006). Without first being decontaminated, the land was taken over by houses, agriculture, and industry. As a result, many residents in Santa Maria lived on top of the oil sumps and were exposed to petroleum waste chemicals. It was not until the turn of the century that cleanup of the sumps was instigated.

2.2 Environmental health concerns

2.2.1 Total petroleum hydrocarbons

Crude oil is made up of several hundred compounds, collectively known as total petroleum hydrocarbon (TPH). The specific composition is based on the geology of the region where the oil was originally excavated (US DHHS, 1999); however, it generally contains benzene, xylene, jet fuel, toluene, and hexane. These compounds can enter the body through inhalation, ingestion, or dermal contact, and have been associated with negatively impacting the blood, immune system, lungs, skin, nervous system, and fetal development (ATSDR, 1999).

Benzene is a known carcinogen, specifically causing acute myeloid leukemia with long-term exposure. Hexane causes peripheral neuropathy, a disorder of the nervous system characterized by numbness or paralysis (US DHHS, 1999). Xylene exposure can affect the kidneys and liver (US DHHS, 1999). Toluene can cause respiratory, liver, and kidney damage (US DHHS, 1999). These are just a few examples of negative health effects specific TPH compounds can have. Because there can be hundreds of chemicals in crude oil and human toxicity data on the majority of the TPH compounds are not available, the actual health effects of crude oil are not known (US DHHS, 1999).

2.2.2 Oil sumps

Dahlgren et al. (2007) showed that homes built on decommissioned oil sumps have higher ambient air and dust concentrations of benzene, xylene, toluene, mercury, and polycyclic aromatic hydrocarbons, compared to homes that were not built on oil sumps. The study also shows that people exposed to oil-contaminated soil are 10 times more likely to develop systemic lupus erythematosus (Dahlgren et al., 2007). Lupus is an autoimmune disease that affects the joints, skin, heart, lungs, brain, and kidneys. Furthermore, Dahlgren et al. (2007) found that people living near oil sumps have higher levels of serum calcium, indicating endocrine malfunction and higher levels of mercury, which negatively affects the immune system.

2.3 Removing total petroleum hydrocarbon from soil

2.3.1 Accelerated solvent extraction (ASE)

This can be used to remove hydrocarbons from soil. The process involves adding a solvent, such as methylene chloride and acetone in a 1:1 ratio, to the soil (Dionex, 2004). Soil is heated to 175°C (350°F) and solvent is added. This method results in an equivalent extraction of contaminants in the soil as other methods, but uses a smaller amount of solvent (Dionex, 2004).

2.3.2 Low-temperature thermal desorption

Low-temperature thermal desorption can be used to remove hydrocarbons from soil, sediments, and sludge. Contaminated soil is removed and transported to a thermal desorption unit. The desorption unit heats the soil and hydrocarbons are volatized. The temperature and the amount of time needed in the unit depends on the contaminant to be removed and the soil conditions. This process creates dust and gas, which is collected for removal without releasing hydrocarbons into the air (Canada Environment, 2002).

2.3.3 Ultrasonic extraction

Ultrasonic extraction uses solvents and a sonicator to remove hydrocarbons from soil. One hundred milliliters of solvent (acetone and n-hexane, in a 1:1 ratio) is used for every 10 grams of soil. After the solvent is applied, a sonicator is used to apply 55 watts for 12 minutes. The soil is then filtered to remove the contaminated residue (Conte et al., 2004).

2.4 Current procedure to remove decommissioned petroleum sumps

The Bureau of Land Management (BLM) issued sump closure guidelines for federal lands that advise but do not guarantee that soil contamination will be completely prevented. The guidelines include the following steps (BLM, 1994):

1. Preliminary work and testing. Remove and legally dispose of all liquids in sumps. Collect at least four soil samples and test for toxins and hazardous waste. Report any excess limits to the BLM, regional water control board, and Federal and State spill authorities to determine next steps.
2. Submit application to the BLM. Components of the report include type of well, location of sump, description of history and sump use, distance from nearest aquifer, crude oil properties, how the contaminated soil will be disposed of and replaced.
3. The BLM evaluates plan and returns a letter with conditions of approval to the operator.
4. The operator then decommissions the oil sump as described in the application and approved by the BLM.
5. During the excavation the operator must submit samples of soil for total petroleum hydrocarbons, general minerals (boron, chloride, sodium, and sulfate), benzene, toluene, ethylbenzene, and xylene.
6. After samples are analyzed and approved, the operator can fill in the sump with clean soil.
7. The operator than submits a final application. The BLM evaluates and accepts the application if no further action is needed.

Although BLM regulations exist for the current decommissioning of oil sumps, there does not appear to be a strong governmental oversight on the remediation of previously decommissioned oil sumps. However, the government will step in when water pollution is identified. An oil sump measuring 500 feet long, 250 feet wide, and 40 feet deep was found in Ventura County and the Aera petroleum company was ordered to remediate the area by the Los Angeles Region Water Quality Control Board after identifying that it is contaminating the nearby Ventura River (CRWQCB, 2000).

2.5 Sump identification and remediation in Santa Maria

The ConocoPhillips and Unocal (now Chevron) corporations are considered responsible for the Santa Maria sump remediation(SBCFD, 2006). Santa Barbara County and the Water Quality Control Board authorized the County Fire Department to regulate remediation, which created a County Site Mitigation Unit (SMU-2) (Doane-Allmon, 2005).

2.5.1 Identification

Sumps are often hidden from plain sight as most of them were covered with soil, grass, houses, and other buildings. ConocoPhillips and Chevron compare past aerial photographs, showing sumps before they were covered, with current aerial photographs to assess the sump placement in terms of residential, commercial, and municipal structures (see Appendix A for an example). Thus far, the primary focus is residential neighborhoods (ConocoPhillips, 2004). Once a potential sump is identified, the soil is tested. If oil-contaminated soil is found, a more thorough investigation is performed to determine the size and perimeter of the sump (Doane-Allmon, 2005).

2.5.2   Remediation

Sumps that have contaminated soil are excavated. Oil-contaminated soil is removed, taken to the Santa Maria landfill, and replaced with clean dirt (Maria). If a sump is beneath a house, ConocoPhillips or Chevron purchases the house, demolishes it, removes the contaminated oil, replaces the soil, and sells the land for new homes (ConocoPhillips, 2004, Goldman, 2006). Homes are purchased for the market price at the time remediation is scheduled (Spencer, 2007). When a sump site has been remediated and replaced with clean dirt "No Further Action" documentation is prepared by SMU-2 as proof of cleanup for current or potential home owners (ConocoPhillips, 2004).

2.5.3   Remediation status

ConocoPhillips identified 70 potential sumps. They have investigated 61 of them and No Further Action documents have been issued for 52 identified sump sites. The nine that still need to be investigated are in commercial or agriculture areas (ConocoPhillips, 2004).

2.5.4   Agricultural land

Previous oil sumps are also being used for agricultural land. It has been discovered that some strawberry fields have been planted on top of abandoned sumps (ConocoPhillips, 2004). Remediation of agriculture fields will occur during the time between harvest and planting, which is often a 1- to 2-month period (Doane-Allmon, 2005).

Growing plants in oil-contaminated soil raises concerns about the quality, safety, and long-term effects of the resulting crops. Currently, oil officials maintain that the remediation is needed to abate a potential nuisance and not for health reasons (Yale, 2008). However, studies show that plants remove TPH chemicals from contaminated soil (Liste and Felgentreu, 2006).

The SMU-2 can determine that a sump does not need remediation. Sumps beneath streets typically are not cleaned since remediation may involve traffic problems and the need to disconnect utilities. The oil companies further argue that since the soil is covered by asphalt, which contains TPH, the covered oil sump does not negatively impact health. Thus far, the County SMU-2 has agreed and has not required remediation for sumps that lie completely under roadways (Doane-Allmon, 2005).

2.6   Lawsuits

In June 2006, 17 residents of Sunrise Hills, a neighborhood in Santa Maria, filed civil lawsuits against Unocal Corporation (now Chevron Corporation), Kerr-McGee Corporation, and ConocoPhillips. Their homes were originally built on decommissioned oils sumps and had to be removed in order for the contaminated soil to be replaced with clean soil. The plaintiffs contend that the corporations contaminated the soil and did not clean it properly before selling the property for the Sunrise Hills residential development. Although the corporations did not admit guilt, settlements were agreed upon, the terms of which are confidential (Yale, 2008).

In November 2006, 18-year-old Scott Chenoweth filed a lawsuit against Unocal, Union Oil, Chevron, ConocoPhillips, Kerr-McGee, and Anadarko Petroleum. Chenoweth, who suffers from acute lymphoblastic leukemia, has lived in Sunrise Hills all his life and claims that oil-contaminated soil was the cause of his disease. Scott seeks compensation for general damages, medical expenses, economic losses, and punitive damages (Spencer, 2006b).

2.7   The current and future Santa Maria

Sump remediation continues in Santa Maria. The focus is now shifting from residential to agriculture and municipal property (Doane-Allmon, 2005).

Recently, the Santa Maria City Council accepted an agreement with the Chevron Corporation to transport of 860,000 cubic yards of oil-contaminated soil from Nipomo-Guadalupe dunes (16 miles from Santa Maria) to Santa Maria (Spencer, 2006a). Chevron is responsible for hydrocarbon release into soil and groundwater in the Guadalupe area (Cuddy, 2006). For $2.5 million compensation, Santa Maria will use the contaminated soil Chevron is required to remove as landfill cover in the Santa Maria landfill (Santa Maria City Hall Council, 2006). The landfill cover aims to reduce rainwater from penetrating the landfill (which already contains contaminated soil extracted from the Santa Maria decommissioned oil sumps) and bringing pollutants into the groundwater below the landfill. The entire transition will require 47,779 truck loads over a period of 2–4 years (County of Santa Maria, 2004), (Cuddy, 2006).

Although the soil cap is meant to prevent pollutants from reaching the groundwater, the contamination in the cap itself has the potential to also pollute the groundwater. In the past, landfills in Canada used hydrocarbon-contaminated soil as a sealing cap. However, this practice was banned and treatment of the contaminated soil has been instigated as it was determined that the hydrocarbons were entering groundwater sources (Canada Environment, 2002).

Why, then, would Santa Maria accept Guadalupe's contaminated soil? The answer to this question lies in Santa Maria's socioeconomic status relative to surrounding communities. Santa Maria is a lower income area. The median household income is \$36,541, 13% below the median income for California (US Census Bureau, 2004). Moreover, 20% of Santa Maria residents live below the federal poverty level (US Census Bureau, 2004).

2.8 Conclusion

The city of Santa Maria has been severely affected by past oil drilling and disposal. Residents of Santa Maria have suffered from negative health effects from the chemicals in the contaminated soil, homes have been destroyed as the contaminated soil is finally being remediated, and agriculture continues to be grown on the decommissioned oil sumps. Although oil companies are starting to clean up the area, the process is a slow one because government does not play a strong enough role in initiating or overseeing the cleanup. Santa Maria's future does not look bright as its current socioeconomic status makes it difficult for the city to resolve its environmental problems.

References

Agency for Toxic Substances and Disease Registry (ATSDR), 1999. ToxFAQ for Total Petroleum Hydrocarbons. Retrieved 5 August 2008, from http://www.atsdr.cdc.gov/tfacts123.html

Bureau of Land Management (BLM), 1994. Oilfield Surface Impoundment Closure Guidelines Bureau of Land Management from http://www.blm.gov/pgdata/etc/medialib/blm/ca/pdf/bakersfield/minerals.Par.69354.File.dat/SumpClosureGuide.pdf

California Department of Conservation, 2007. 2006 Annual Report of the State Oil and Gas Supervisor. Sacramento, CA.

California Regional Water Quality Control Board (CRWQCB), 2000. To Clean Up and Abate Conditions of Water Pollution Caused by the Release of Industrial Waste From an Oilfield Waste Sump in Ventura, CA. State of California.

Canada Environment, 2002. TAB#13 Soil Remediation: Low Temperature Thermal Desorption. TABs on Contaminated Sites. Retrieved 1 August 2008, from http://www.on.ec.gc.ca/pollution/ecnpd/tabs/tab13-e.html

ConocoPhillips, 2004. Santa Maria Valley Sumps Program. Retrieved 1 August 2008, from http://smvsumps.com/contact.html

Conte, P. et al., 2004. Soil Remediation: Humic Acids as Natural Surfactants in the Washings of Highly Contaminated Soils. Environmental Pollution 135, 515–522.

County of Santa Maria, 2004. Fact Sheet and Frequently Asked Questions: Non-Hazardous Hydrocarbon Impacted Soil. Retrieved 22 July 2008, from http://cityofsantamarianhis.info/downloads/nhis-facts.pdf

Cuddy, B., 2006. Guadalupe Dunes Cleanup to Begin. The Tribune, 8 February.

Dahlgren, J., Takhar, H., Anderson-Mahoney, P., Kotlerman, J., Tarr, J., Warshaw, R., 2007. Cluster of Systemic Lupus Erythematosus (SLE) Associated with an Oil Field Waste Site: A Cross Sectional Study. Environmental Health 8 (6).

Dionex, 2004. Extraction of Total Petroleum Hydrocarbon Contaminats (Diesel and Waste Oil) in Soils by Accelerated Solvent Extraction (ASE). Sunnyvale, CA.

Doane-Allmon, J., Boyd, H. (2005, October). Drilling Sump Restoration in Santa Maria Valley, California. RemTech 2005 - Remediation Technologies Symposium, Alberta, Canada.

Goldman, M., 2006. Notice of Public Review and Request for Comments, Re: Sunflower Court, Santa Maria Contaminated Soil Cleanup Project.

Liste, H-H., Felgentreu, D., 2006. Crop Growth, Culturable Bacteria, and Degradation of Petrol Hydrocarbons (PHCs) in a Long-term Contaminated Field Soil. Applied Soil Ecology 31 (2), 43–52.

Santa Barbara County Fire Department (SBCDF), 2006. Oilfield/Lease Decommissioning and Restoration Program. Retrieved 5 August 2008, from http://www.sbcfire.com/hm/programs/smu2faq.html

Santa Maria City Hall Council, 2006. California Regular Meeting, Santa Maria.

Spencer, M., 2006a. $900,000 Unocal Deal Moves Ahead. Santa Maria Times, 14 June.

Spencer, M., 2006b. Local Teen Files Lawsuit Over Cancer. Santa Maria Times, 3 November.

Spencer, M., 2007. Homes May Be Demolished to Clean Up Soil. Santa Maria Times, 29 April.

US Census Bureau, 2004. State and County QuickFacts, Santa Maria, California, from http://quickfacts.census.gov/qfd/states/06000.html

US Department of Health and Human Services (US DHHS), 1999. Toxicological Profile for Total Petroleum Hydrocarbons. Atlanta, GA.

Yale, S., 2008. Residents Settle Suits Against Oil Companies. Santa Maria Times, 26 February.

3 The Santa Barbara oil spill of 1969

3.1 Introduction

This chapter provides a case study of a major oil spill. A review of this case study provides some lessons and highlights poor environmental management practices that should be avoided.

3.2 The incident

In 1969, a Union Oil of California drilling rig called Platform A (or alpha), located offshore of Santa Barbara, was extracting pipe from a 3500-foot-deep well. There was a pressure difference created by the extraction of the pipe that was not sufficiently compensated for by the pumping of drilling mud back into the well. This created a pressure increase that, despite attempts to cap the well, resulted in extreme pressure below the ocean floor and the bursting of natural gas from the hole. The intensity of the release caused five cracks in the sea floor around the drill casings along an east–west fault, releasing a large volume of oil and natural gas from deep beneath the earth (Clarke and Hemphill, 2002; GOO, 2008, SBWCN, 2008). Over the next 11 days, and depending on the reporting source, there was anywhere from 200,000 (SBWCN, 2008) to 3 million gallons (Clarke and Hemphill, 2002) of crude oil spread over 800 square miles. Thick tar coated beaches from Rincon Point to Goleta, damaging 35 miles of coastline. The slick moved south and affected Anacapa Island's Frenchy's Cove, as well as beaches on Santa Cruz, Santa Rosa, and San Miguel islands (Clarke and Hemphill, 2002).

3.3 Who was responsible for the accident?

Two main players were acknowledged to be responsible for the 1969 oil spill: Union Oil of California and the United States Department of the interior, specifically the US Geological Survey (USGS). The USGS had already specified a particular length of casing on the pipe used by Union Oil on Platform A. A casing is a safety device that reinforces the well to prevent blowouts. Union Oil was granted a waiver on the length of the casing for this well and used a shorter casing that was below federal and State of California guidelines (Clarke and Hemphill, 2002). Donald Solanas of the USGS approved the waiver. However:

On wells as deep as A-21 [note: the well designation for Platform A], federal regulations called for a standard minimum of 300 feet of conductor casing – the first string of protective casing normally set beneath the ocean floor – and a blowout-prevention device atop it. Similarly, these regulations called for approximately 870 feet of surface casing – a secondary string set to greater depth and generally installed when exploratory operations suggest the presence of a high-pressure gas-zone. Yet Solanas – exercising his legitimate statutory discretion – had authorized Union to drill A-21 without installing any surface casing at all. Moreover, he had permitted Union to run its conductor casing down to only 238 feet beneath the ocean floor.

(Nash et al., 1972)

Thus, when the well had the blowout; even though it was capped it still resulted in fragmentation of the wellhead. Investigators would later determine that more steel pipe sheeting inside the drilling hole would have prevented the rupture (SBWCN, 2008).

3.4 Federal response

The Federal response to the blowout included an immediate halt to drilling on the outer continental shelf (OCS) and review of the federal drilling regulations, ordered by the Secretary of the Interior, Walter Hickel. Additionally, he ordered a 34,000-acre buffer zone seaward of the existing 21,000-acre federal ecological preserve between Summerland and Coal Oil Point. In 1970, Federal drilling activities were allowed to continue on the OCS. This was allowed only with stricter regulations. Future OCS oil and gas leasing, as well as leasing in State waters, would require a formalized environmental public review process under the newly enacted National Environmental Policy Act and the California Environmental Quality Act (CSBPD, 2005).

3.5 Lawsuits

The immediate human impact was to the tourism industry in California, as well as owners of beachfront homes, apartments, hotels, and motels. A class-action lawsuit filed against Union Oil of California awarded nearly $6.5 million to owners of these homes and businesses. Additionally, recreational boat owners and commercial boat owners and nautical suppliers received $1.3 million for property damage and loss of revenue. While some commercial fishers lost access to some fisheries temporarily, no lawsuit seems to have been awarded (CSBPD, 2005).

The State of California, County of Santa Barbara, and the Cities of Santa Barbara and Carpinteria filed and eventually settled a lawsuit with Union Oil for loss of property in the amount of $9.5 million (Welsh, 1989).

While the tourist industry suffered in 1969, it appeared to recover in subsequent years.

3.6 Crude oil and the ocean environment

Crude oil is a mixture of thousands of different hydrocarbons (Patin, 2008). A marine ecosystem destroys, metabolizes, and deposits the excessive amounts of hydrocarbons, transforming them into more common and safer substances. The ecosystem deals with the oil in a number of ways: physical transport is the thinning, spreading, travel, and dispersing of the oil over the ocean surface and over large distances. Dissolution is the action of the component chemicals dissolving into the water. Emulsification can occur either as water in oil or as oil in water, and can create a 'chocolate mousse', lasting in the marine environment for over 100 days. Oxidation occurs after about a day of the oil being in the ocean environment. As components oxidize they form other chemicals that are more water soluble and more toxic. Sedimentation occurs as the oil sticks to particles in the ocean water, which then fall to the ocean floor. Biosedimentation also occurs as plankton and other organisms absorb the emulsified oils, which they then process and excrete. These excretions then settle to the ocean floor. These heavy concentrations can exist for months or years in the soils of the ocean floor (Patin, 2008).

The fate of most petroleum substances in the marine environment is ultimately defined by their transformation and degradation due to microbial activity (Patin, 2008). In microbial degradation, or the processing of the oils by microorganisms, a number of bacteria and fungi are able to feed off the hydrocarbons. The degree and rate of hydrocarbon biodegradation depends on a number of factors, including composition of the oil, temperatures, concentrations, and the number of microorganisms in the area. Aggregation, of the formation of petroleum lumps or clumps of tar, occurs as the lighter substances in oil are otherwise dispersed. Oil aggregates can exist from a month to a year in the enclosed seas and up to several years in the open ocean. They can end up slowly degrading in the ocean, on the shore if they are washed there by currents, or on the sea bottom if they lose their floating ability. Amazingly, if the toxic load of the oil on the environment does not surpass critical limits, the ocean will eventually undergo a self-purification of the oils, which eventually degrade to carbon dioxide and water (Patin, 2008).

3.7 Specific ecological impact of the Santa Barbara oil spill

Oil along the coast of Santa Barbara was at some points 6 inches thick, muting the waves on the beach and producing a stench of petroleum described as 'inescapable' (SBWCN, 2008). Marine animals were coated with oil, while others ingested it, resulting in poisoning and suffocation. Blow holes of dolphins were clogged, causing massive lung hemorrhages. Animals that ingested the oils, such as seals, were poisoned. Birds that feed by diving into the water became soaked with tar. Other animals survived by avoiding the area, such as gray whales migrating to their calving and breeding grounds in Baja California, or shorebirds like plovers, godwits, and willets, which feed on sand creatures (Clarke and Hemphill, 2002).

An estimated 3686 birds died because of contact with oil. Follow-up aerial surveys, taken 1 year later, found only 200 grebes in an area that had previously drawn 4000–7000 (SBWCN, 2008). The spilled oil killed innumerable fish and intertidal invertebrates, devastated kelp forests, and displaced many populations of endangered birds (Clarke and Hemphill, 2002).

3.8   Cleanup efforts

The response by the community was immediate. People of all ages, groups, and political viewpoints volunteered to become a part of the cleanup.

3.8.1   Beaches and oceans

Beach cleanup included distributing piles of straw to absorb oil that washed onshore. Beach sand contaminated by the oil was bulldozed into piles and then transported away. On the ocean, skimmer ships gathered oil from the surface. Detergents were spread over the slick to disperse the oil (SBWCN, 2008).

3.8.2   Wildlife

Three emergency bird treatment centers were set up during the crisis, including the Santa Barbara Zoo. Volunteers plucked oiled birds from local beaches. Birds were bathed in Polycomplex A-11, medicated, and placed under heat lamps to stave off pneumonia. While these were noble efforts, the survival rate was less than 30% for birds that were treated. Ironically, birds continued to die on the beaches due to the detergents used to disperse the oil slick. The chemicals stripped their feathers of the natural waterproofing used to keep them afloat (Clarke and Hemphill, 2002). According to the County of Santa Barbara Planning and Development, all species recovered after a few years (CSBPD, 2005).

3.9   Impact on legislation and regulations

As noted above, offshore drilling was halted for a period of time, regulations were reviewed, and drilling was again allowed to continue. Future OCS oil and gas leasing, as well as leasing in State waters, would require a formalized environmental public review process under the newly enacted environmental laws (CSBPD, 2005).

The *National Environmental Policy Act (NEPA)* was signed into law on 1 January 1970. It applies to all federal agencies and most of the activities they manage, regulate, or fund that affect the environment. It requires all agencies to disclose and consider the environmental implications of their proposed actions (CSBPD, 2005).

The *California Environmental Quality Act (CEQA)* was enacted: (1) to inform government decision-makers and the public about the potential environmental effects of proposed activities; (2) to identify ways that a proposed project's environmental damage can be avoided or significantly reduced; (3) to prevent significant, avoidable damage by requiring changes in projects, either by the adoption of alternatives or imposition of mitigation measures; and (4) to disclose to the public why a project was approved if that project would have significant environmental effects (CSBPD, 2005). The CEQA applies to all governmental agencies at all levels in California, although it does not apply to the California legislature. It affects the approval of projects subject to the CEQA that may result in one or more significant effects on the environment, effectively requiring those responsible for detrimental environmental effects to mitigate those effects through feasible alternatives.

The *California Coastal Commission* was established by voter initiative in 1972 (Proposition 20) and later made permanent by the Legislature through adoption of the California Coastal Act of 1976 (CCC, 2008). This commission today has powerful control over human activities that impact California's coastal areas (Clarke and Hemphill, 2002).

The *State Land Commission* of California banned offshore drilling for 16 years, until the Reagan Administration took office (Clarke and Hemphill, 2002).

President Nixon signed the *National Environmental Policy Act of 1969*. This act was the precursor for the creation of the July 1970 establishment of the *Environmental Protection Agency* (Clarke and Hemphill, 2002; SBWCN, 2008).

3.10 Conclusion

With increased regulations and safety measures, oil spills still occur. There were 37 documented spills nationally in 1994, many of them far exceeding that of the January 1969 spill in magnitude and consequences (Clarke and Hemphill, 2002). As of September 2008, 19 platforms were drilling off Santa Barbara's outer continental shelf (CSBPD, 2005). In the past 38 years, there have been 1000 barrels spilled from these platforms (SOS, 2008). Although the initial response to the Santa Barbara oil spill was heightened environmental awareness and increased litigation, it carried little longevity in terms of preventing future oil spills. Therefore, it is up to the individual oil company to ensure that their environmental impact when drilling oil is as low as possible.

References

California Coastal Commission (CCC), 2008. Program Overview. Available at http://www.coastal.ca.gov/whoweare.html.
Clarke, K.C., Hemphill, J.J., 2002. The Santa Barbara Oil Spill, A Retrospective. In: Danta, D. (Ed.), Yearbook of the Association of Pacific Coast Geographers. University of Hawaii Press, Vol. 64, pp. 157–162.

County of Santa Barbara Planning and Development (CSBPD), 2005. Blowout at Union Oil's Platform A. Energy Division. Available at http://www.countyofsb.org/ENERGY/information/1969blowout.asp.

GOO: Get Oil Out!, 2008. GOO – History. Available at http://www.getoilout.org/about.html.

Nash, A.E., Mann, D.E. and Olsen, P.G., 1972. Oil Pollution and the Public Interest: A Study of the Santa Barbara Oil Spill. UC Institute of Governmental Studies, Berkeley. Available at http://www.willjohnston.com/articles2000/evos/evos_plus.doc, 2.

Patin, S., 2008. Oil Spills in the Marine Environment (Cascio, E., translator). Available at http://www.offshore-environment.com/oil.html.

Santa Barbara Wildlife Care Network (SBWCN), 2008. Santa Barbara's 1969 Oil Spill. Available at http://www.sbwcn.org/spill.shtml.

SOS California, 2008. SOS's Solution: Education. Available at http://www.soscalifornia.org/solution.html.

University of California, Santa Barbara (UCSB), 2008. Environmental Studies at UCSB. Available at http://www.es.ucsb.edu/general_info/.

4 Exxon Valdez oil spill

4.1   Introduction

Poor management practices have resulted in numerous disasters during crude oil transport operations. One case study presented in this chapter is the Exxon Valdez oil spill.

4.2   The event

On 24 March 1989, the Exxon Valdez ran aground on a large but newly formed ice shelf in Prince William Sound. The accident spilled 11 million gallons of oil into the ocean. Over the next 3 days, oil spread across 1300 miles from Alaska to northern Washington, resulting in widespread damage to the ecosystem. The waves washed the oil 120 feet up the berm to upset inland habitats and seep into the gravel-covered beaches. In the wake of the disaster, 250,000 seabirds, 2800 otters, 300 harbor seals, 250 bald eagles, and countless species of fish washed up dead on the affected beaches. The world's largest estuary would remain permanently damaged by the technological disaster (Grabowski and Roberts, 1996; Paine et al., 1996; Picou, 2000; Carson et al., 2008).

Serious human error resulted in the accident. The captain of the Exxon Valdez was a known alcoholic currently undergoing treatment. However, no monitoring was performed to ensure that the captain remained sober during his time at the helm. The captain's poor judgment and alcohol intake impaired his ability to react to changing conditions in the volatile Arctic climate (Paine et al., 1996).

Despite the obvious error, Exxon withheld admission of guilt. For 3 days, the oil spilled while the government, local industry, and Exxon battled over responsibility. Local fishermen were eager to salvage what was possible of their livelihood but Exxon feared liability issues related to non-industry cleanup. The heightened awareness of future fines and liability lessened cooperation and postponed intervention until government order initiated the process. Exxon began the rigorous cleanup process under the watchful eye of government and environmentalists. The delay proved costly. The resultant damage cost \$5 billion in lost passive use, over \$3 billion in cleanup, and \$287 million in lost wages to the local fishing industry (Paine et al., 1996; Duffield, 1997; Hayes and Michel, 1999a; Picou, 2000; Carson et al., 2008).

The Exxon response generated critical media. Nonstop coverage spread news of the Valdez and its mother company around the world. Exxon provided ample news stories and the CEO appeared 'arrogant' and 'silly' to the world. Among the blunders committed by the company in the public eye were:

- Exxon officials failed to show up on-site or issue public statements.
- Exxon failed to train recovery procedures or adhere to guidelines for Alaskan pipeline use.
- Exxon argued over responsibility and delayed cleanup efforts.
- Exxon refused local efforts and the input of environmental groups.
- Exxon consistently understated the extent of damage.
- Exxon reported useless data such as number of boats on-site or number of oil barrels recovered and doctored reports to improve the on-paper data.
- The Exxon CEO made silly and selfish statements that implied the public furor was a witch hunt or that the public would only be happy 'if he put a gun to his head and pulled the trigger'.
- Exxon released a disturbing ad comparing the environmental harm in Alaska to Marilyn Monroe's mole and stating that the loss was similar to a mole removal and hardly marred Alaska's beauty.

The blunders resulted in a public image nightmare as the media portrayed Exxon as 'arrogant ... ruthlessly capitalistic ... and cold and calculating'. This image would hurt Exxon in the resultant civil court cases (Tyler, 1992; Polinsky and Shavell, 1994; Paine et al., 1996; Duffield, 1997).

4.3   The environment

The crude oil spilled into the bay and covered the water in a smooth slick. While crude oil is not immediately toxic, the oil slick immersed the habitat in a viscous, non-aqueous solution. Fish, otters, sea birds, and other marine life became drenched in the slick, sticky substance. The oil caused sea life to lose buoyancy, resulting in drowning for mammalian species and fatigue for underwater species. Seabirds could not fly. Mussels, algae, and other mollusks died of suffocation (Carson et al., 2008).

Upon exposure to air, water, and bacteria, the crude oil rapidly began to degrade into harmful by-products. Toxic chemicals found in the water and soil included sulfides, sulfur, benzene, phenanthrene, naphthalene, dioxygenase, and polycyclic aromatic compounds. These chemicals may decrease tissue oxygenation, depress central nervous system function, and induce liver, lung, eye and reproductive abnormalities. Several by-products, in particular benzene, are proven carcinogens (Mann et al., 1999; Sullivan and Krieger, 2001; Samanta et al., 2002; Eyong et al., 2004; DHHS website, http://www.atsdr.cdc.gov/tfacts69.html).

Animals and humans in the environment were exposed to the chemicals through direct contact, inhalation, and the food supply. Animal communities were demonstrated to show long-term impacts that have not entirely recovered after 20 years. Socially organized animals seemed most affected, with significant changes in social patterns, caregiving, and nesting behavior. Fish populations showed changes in schooling behavior and reproductive abnormalities over several generations (Sullivan and Krieger, 2001; Peterson et al., 2003).

Human impact was not as carefully documented as damage to the ecosystem. Human exposure was limited by rapid response to cease ingestion of marred animals or polluted water as well as limitations on exposure to the chemicals during cleaning. However, documented reports indicate that depression and social disruption also increased significantly in the wake of the Exxon tragedy.

Technological disasters have been shown to result in higher levels of post-traumatic stress disorder than found with natural disasters. The prevalence of post-traumatic stress disorder in Alaska increased to 17.2%, a rate three times the rate seen after the Mount St Helens eruption. Community conflict tends to increase the experience of post-traumatic stress disorder. The arguments, delay in cleaning, and lengthy disruption of the fishing industry heightened the stress response among the Alaskan citizens. Furthermore, native Alaskans continue to rely on subsistence food gathering for survival. The Exxon Valdez limited access to wildlife, so native Alaskans were unable to find food. Fears over the contamination in the water and food supply heightened tensions in an isolated environment with minimal access to outside resources. Exxon's attempts to limit responsibility created further distrust and concern as Alaskans began to doubt the effectiveness of cleanup efforts (Palinkas et al., 1993; Lanier et al., 1996; Mann et al., 1999; Morita et al., 1999; Arata et al., 2000; Picou, 2000; Sullivan and Krieger, 2001; Eyong et al., 2004; DHHS website, http://www. atsdr.cdc.gov/tfacts69.html).

Although reported data merely indicate a change in the psychosocial profile of Alaska, other oil spills in various countries have reported significant physical problems related to oil by-product exposure. A review of Alaska's health problems for the 10-year period following the spill indicates an increase in many health problems, including a 25% increase in cancer rates. However, causality has not been proven. Increased attention to health problems of native Alaskans may have influenced diagnostic procedures and changed attendance at local clinics during the same period. As such, physical health changes cannot be directly attributed to the Exxon Valdez disaster (Palinkas et al., 1993; Lanier et al., 1996; Sullivan and Krieger, 2001).

4.4 Cleanup processes

In the immediate aftermath of the Exxon spill, the oil company attempted to burn the oil off the surface of the water. The technique was highly successful and resulted in a removal of 40% of the oil at the burn site. However, the native Alaskan population was highly upset by the 'burning sea' and local citizens grew concerned over the fumes. The Coast Guard ordered the burns stopped. Exxon was forced to resort to less effective methods of oil removal (Paine et al., 1996).

The varied terrain resulted in difficult cleanup. Oil degraded at varying rates dependent on the topography of the area. Hot-water and high-pressure washing of the beaches, water skimming, and bioremediation were all utilized in an attempt to remove the harmful chemicals from the water and sands in the spill area. Each technique proved minimally successful and most areas required more than one cleaning in the 10 years following the spill (Pritchard et al., 1992; Paine et al., 1996; Hayes and Michel, 1999b; Samanta et al., 2002; Peterson et al., 2003).

Several of the cleaning techniques used proved more harmful than beneficial. Washing was discouraged by environmental groups as the high-pressure, hot-water sprays damage remaining algae and mussel communities. However, the pressure on Exxon to clean the beaches led to a preference for quick results. Skimming caused more harm than good as well. Skimming is performed by large boats pulling absorbent nets across the water. The process is inefficient, removed only 10–15% of surface oil, and 40 times as much oil was used as spilled into the harbor (Paine et al., 1996; Peterson et al., 2003).

The process of bioremediation was improved in order to hasten hydrocarbon degradation. Seeding the oil with Inipol and other nitrogen-based fertilizers resulted in faster breakdown of oil by-products. Highly porous beaches with high sand–water interfaces for maximum oil emulsion along the Prince William Sound were particularly conducive to bioremediation (Pritchard et al., 1992; Bragg et al., 1994).

After 1 year, very little visible oil remained on the beaches. The lack of visible oil perpetuated the belief that the cleanup process had been completely successful and thoroughly accomplished. However, gravel on the beaches trapped the oil into lower levels of sediment. Four years after the spill, oil was present along 7 km of shoreline in 100 distinct locations. Five years after the spill, 13% of oil remained in the sediment. Ten years later, eight sections of beach were cleaned after oil began to surface. The ecosystem continues to show signs of damage and several beaches still test positive for oil by-products (Peterson et al., 2003).

Exxon reports indicated that $2 billion was spent on the initial cleanup efforts. Another $300 million was provided to offset lost wages for local fishermen. Subsequent costs for damage assessment and restoration totaled $3.2 billion. The spill has been one of the costliest technological disasters in history (Paine et al., 1996).

4.5 Environmental justice

Under the Code of Federal Regulations, oil companies are required to pay for damages if the oil-related injury directly results from company actions. Oil-related injury is defined as 'any measurable, adverse change, either short or long term, in the chemical or physical quality of the natural resource.' Oil companies are required to pay for 'return to baseline' or, in other words, a complete restoration of the ecosystem that would have existed had the injury not occurred (Paine et al., 1996).

The language of the code is vague enough to leave ample room for interpretation. Exxon interpreted the code to define recovery as the establishment of a healthy, biologic ecosystem but not necessarily the ecosystem present at the time of damage. Exxon scientists cited several studies that demonstrated the volatility of Alaskan ecosystems, including the effect of harsh weather and unpredictable temperatures on the animal and plant communities over the past years. Per Exxon, natural variability meant that it was impossible to know what conditions would have been like if the spill had not occurred. Exxon has produced scientific papers, hired scientists, and encouraged research to support its position that Alaska experienced minimal harm from the spill. As such, Exxon limited its involvement in restoration (Paine et al., 1996; Wiens et al., 1996).

Despite Exxon's protestations, hundreds of civil lawsuits were filed on behalf of private and public interests over the course of 20 years. The original jury found Exxon liable for $5 billion while Exxon settled the native subsistence case for $20 million. However, the Supreme Court reduced that claim to one-fifth the original amount and permitted Exxon to claim cleanup effort costs as payment towards the fine (Brooks, 2002).

Figuring in the Supreme Court's decision was the likelihood that media involvement had led the jury to ask for exorbitant fines not based on sound scientific data. Exxon's actions immediately following the oil spill had infuriated the public and created harsh public sentiment demanding maximum punishment. The media coverage broadcasted Exxon's many mistakes and likely prejudiced the jury towards harsh punishments and excessive fines (Tyler, 1992; Polinsky and Shavell, 1994; Paine et al., 1996; Duffield, 1997; Hastie et al., 1998; Carson et al., 2008).

Furthermore, estimations of damage were based on analyses that were easily contested. The original jury depended on data from a contingency valuation study. This study factored in not only the estimated cost for cleanup and restoration but an estimate for the passive loss value. This number was derived from a survey of residents in various US locations to discover what personal monetary investment would be offered to restore Alaska. The purpose was to determine the monetary worth of the destroyed areas. The median response was $31 per household. As the area destroyed was public land, the contingency value was determined assuming that all Americans would pay the determined amount. As such, the amount of damage was assessed to be astronomically large. The Supreme Court rejected this value in favor of less extreme values based on market values and costs of restoration (Paine et al., 1996; Hastie et al., 1998; Carson et al., 2008).

4.6 Government response

The impact of Exxon Valdez extended beyond the courts. In 1990, the US legislature passed new legislation meant to prevent further technological disasters related to oil production, shipment, and refinery. The act called for maximum responsibility and increased corporate responsibility. However, enforcement of the act lasted only as long as public outcry over Exxon Valdez continued (Brooks, 2002).

References

Arata, C.M., et al., 2000. Coping with Technological Disaster: An Application of the Conservation of Resources Model to the Exxon Valdez Oil Spill. Journal of Traumatic Stress 13 (1), 23–39.

Bragg, J.R., et al., 1994. Effectiveness of Bioremediation for the Exxon Valdez Oil Spill. Nature 368, 413–418.

Brooks, R.W. 2002. Liability and Organizational Choice. Journal of Law and Economics 45, 91–125.

Carson, R.T. et al., 2008. A Contingent Valuation Study of Lost Passive Use Values Resulting from the Exxon Valdez Oil Spill. MPRE Paper 6984, February. Available at http://mpra.ub.uni-muenchen.de/6984/.

DHHS. Agency for Toxic Substances and Disease Registry. ToxFAQs. As seen at http://www.atsdr.cdc.gov/tfacts69.html.

Duffield, J., 1997. Nonmarket Valuation and the Courts: The Case of the Exxon Valdez. Contemporary Economic Policy 15, 98–110.

Eyong, E.U., et al., 2004. Haematoxic Effects Following Ingestion of Nigerian Crude Oil and Crude Oil Polluted Shellfish by Rats. Nigerian Journal of Physiological Sciences 19 (1–2), 1–6.

Grabowski, M., Roberts, K.H., 1996. Human and Organizational Error in Large Scale Systems. IEEE Transactions on Systems, Man, and Cybernetics – Part A: Systems and Humans 26 (1), 2–15.

Hastie, R., Schkade, D.A., Payne, J.W., 1998. A Study of Juror and Jury Judgments in Civil Cases: Deciding Liability for Punitive Damages. Law and Human Behavior 22 (3), 287–314.

Hayes, M.O., Michel, J., 1999a. Factors Determining the Long-term Persistence of Exxon Valdez Oil in Gravel Beaches. Marine Pollution Bulletin 38 (2), 92–101.

Hayes, M.O., Michel, J., 1999b. Weathering Patterns of Oil Residues Eight Years After the Exxon Valdez Oil Spill. Marine Pollution Bulletin 38 (10), 855–863.

Lanier, A.P., Kelly, J.J., Smith, B., Harpster, A.P., Tanttila, H., Amadon, C., Beckworth, D., Key, C., Davidson, A.M., 1996. Alaska Native Cancer Update: Incidence Rates 1989–1993. Cancer Epidemiology Biomarkers and Prevention 5 (9), 749–751.

Mann, J.M., Gruskin, S., Grodin, M.A., 1999. Health and Human Rights: A Reader. Routledge. 1999.

Morita, A., et al., 1999. Acute Health Problems among the People Engaged in the Cleanup of the Nakhodka Oil Spill. Environmental Research 81 (3), 185–194.

Paine, R.T., et al., 1996. Trouble on Oiled Waters: Lessons from the Exxon Valdez Oil Spill. Annual Review of Ecological Systems 27, 197–235.

Palinkas, L.A., et al., 1993. Community Patterns of Psychiatric Disorders after the Exxon Valdez Oil Spill. American Journal of Psychiatry 150, 1517–1523.

Peterson, C.H., et al., 2003. Long-term Ecosystem Response to the Exxon Valdez Oil Spill. Science 302, 2082–2086.

Picou, J.S., 2000. The "Talking Circle" as a Sociological Practice: Cultural Transformation of Chronic Disaster Impacts. Sociological Practice: A Journal of Clinical and Applied Sociology 2 (2), 77–97.

Polinsky, A.M., Shavell, S., 1994. A Note on Optimal Cleanup and Liability after Environmentally Harmful Discharges. Research in Law and Economics 16, 17–24.

Pritchard, P.H., et al., 1992. Oil Spill Bioremediation: Experiences, Lessons and Results from the Exxon Valdez Oil Spill in Alaska. Biodegradation 3, 315–335.

Samanta, S.K., Singh, O.V., Jain, R.K., 2002. Polycyclic Aromatic Hydrocarbons: Environmental Pollution and Bioremediation. Trends in Biotechnology 20 (6), 243–246.

Sullivan, J.B., Krieger, G.R., 2001. Clinical Environmental Health and Toxic Exposures. Lippincott Williams & Wilkins.

Tyler, L., 1992. Ecological Disaster and Rhetorical Response: Exxon's Communications in the Wake of the Valdez Spill. Journal of Business and Technical Communication 6, 149–171.

Wiens, J.A., et al., 1996. Effects of the Exxon Valdez Oil Spill on Marine Bird Communities in Prince William Sound. Ecological Applications 6 (3), 828–841.

5 Best practices for developing fugitive emissions inventories

5.1 Introduction

A Congressional report prepared under Representative Waxman's leadership showed that for valves alone the average oil refinery under-reports fugitive emissions by nearly a factor of 5 (Waxman report, 1999). Another study by the Environmental Integrity Group (EIG) noted that the US Environmental Protection Agency (EPA) and state governments appear to be under-reporting refinery toxic air emissions substantially (http://www.commondreams.org/news2004/0622-14.htm). The analysis finds that the presence of the carcinogens benzene and butadiene in the air in the USA may be four to five times higher than the level the EPA reports to the public through the Toxic Release Inventory (TRI) program. The study, which is based on findings by the Texas Commission on Environmental Quality (TCEQ), applies the Commission's findings on the under-reporting of certain toxic emissions nationwide and concludes that at least 16% of toxic air emissions from all sources are not typically reported.

The EIG study states:

> ... EPA has failed to improve monitoring and reporting of toxic air pollution. The investigators charge that the EPA has moved in the opposite direction and has weakened some federal monitoring requirements ... EPA adopted new rules that actually weakened air emission reporting requirements ... EPA's old rules required that major air pollution sources conduct monitoring sufficient to reveal whether or not the source was complying with federal pollution limits ... EPA revised these rules to only require monitoring that occurs more than once every 5 years. Such infrequent monitoring is clearly inadequate for tracking compliance and means that more sources will be using emission calculations and estimations, rather than actual monitoring, to report emissions.

The above studies highlight that under-reporting happens because most air pollution is now estimated and not actually monitored in the USA. To make matters worse, the estimates are prepared by the facilities that generate pollution. This is a conflict of interest because such facilities have financial incentives to keep emissions reporting figures low. Refineries report their toxic emissions under an honor system that is based on calculations and, as described in this chapter, these methods are not reliable.

Studies by the Texas Commission on Environmental Quality have quantified the extent to which refineries in Texas under-report certain toxic emissions. The study reveals that, if Texas's results are applied nationwide, refineries and chemical plants failed to report at least 330 million pounds of toxic hydrocarbon emissions, including known carcinogens like benzene. Texas officials limited their research to certain hydrocarbons believed to play a major role in causing rapid ozone formation in the Houston area. Of these, 10 hydrocarbons – ethylene, toluene, hexane, xylene, propylene, styrene, benzene, cyclohexane, ethylbenzene and butadiene – are chemicals that are reported to the TRI. In their report, the 2001 TRI levels reported for chemical plant and refinery emissions of those 10 hydrocarbons were adjusted based on the methodology developed by Texas. Emissions were adjusted for only chemical plants and refineries in four Standard Industrial Codes (SICs), which is the basis for claiming that under-reporting is a national problem.

We note several examples of under-reporting by various companies in the following paragraphs. Our aim in describing these cases is to point out how under-reporting practices extend beyond the reliance on calculation methods that have not been validated.

In a civil suit against Plains Exploration & Production Company (PXP) it was learned that the facility operator did not maintain formal records of leaks and fugitive VOC emissions including benzene (*Hensley* v. *Hoss*). In this case, an independent auditing service was used to monitor sources of emissions twice a week. This service not only repairs any leaking valves and other piping components, but is supposed to maintain an accurate accounting of the leaks for quarterly reporting purposes. Another consultant was engaged to prepare quarterly emissions reports based on the monitoring reports provided by the auditing service and submitted these for several years to the California Air Quality Management Board. Unbeknownst to either the monitoring service or the consultant responsible for preparing the emissions reports, PXP had its maintenance crews independently identify and repair leaks and spills from various process equipment and piping components. The maintenance crews were not equipped with instrumentation, but rather used sight, smell, and hearing to detect gaseous emissions and hydrocarbon leaks from its oil and gas field, and, we believe, its gas processing plant. While internal maintenance records were maintained of the repairs, no records of the numbers and concentrations of the emissions were ever documented. PXP's personnel did not follow EPA monitoring protocols because they were never provided with any monitoring instrumentation.

The Citgo Petroleum Corporation (CITGO) Lake Charles Manufacturing Complex is the fourth largest refinery in the USA. On 19 June 2006 a heavy rain event occurred. Nearly 99,000 barrels (4.16 million gallons) of waste oil were released into the Calcasieu River from the CITGO refinery near Sulphur, LA. The spill occurred in part due to heavy rains overwhelming storage tanks at the refinery. Of this catastrophic release, 53,000 barrels (2.23 million gallons) of the oil and 259,524 barrels (10.9 million gallons) of contaminated wastewater were released to the Indian Marais over a 2-day period, with approximately 25,000 barrels (1 million gallons) of the waste oil having migrated into the Calcasieu River over the same release period. More than 50% of the spill was not contained and was released into the surrounding area. In an ongoing litigation against this company, it has been learned that not a single pound of this release was ever reported in the annual TRI. It was also learned that CITGO relied on fugitive air monitoring practices in which it never calibrated field instruments, and hence provided inaccurate information to both cleanup crews and the community on their risks from exposures to fugitive air emissions.

A study by the Alberta Research Council[1] that investigated the plume of contaminants emanating from a Canadian oil refinery using high-tech sniffing equipment found the facility dramatically underestimated its releases of dangerous air pollutants. The subject refinery was found to release 19 times more cancer-causing benzene than it reported under Environment Canada disclosure regulations, about 15 times more smog causing volatile organic compounds, and nine times more methane, a greenhouse gas, according to the study. This testing is believed to be the first at a North American refinery using the sophisticated technology relying on lasers, and is considered state of the art; but this same technology, developed by British Petroleum, has been in widespread use in Europe for nearly two decades. The investigators go on to report that, based on the study, funded by the federal Alberta and Ontario governments, it is likely that all refineries in Canada and the USA are undercounting emissions because they follow an estimating protocol developed by the American Petroleum Institute and the US EPA. Under the US protocol, refineries do not calculate their actual emissions, but try to reach approximate figures using technical assumptions and mathematical equations. The study notes that estimating refinery pollution is difficult because most emissions do not come from easy-to-monitor sources, such as smokestacks, but are wafting from numerous leaking valves, storage tanks, drains, and vents. The same may be said for many other industry sectors, including gas processing and natural gas-gathering fields. These fugitive emissions are estimated under the API–EPA protocol, which for the petroleum refining sector have not been revised since 1972. The study, which tabulated actual emissions in the air around the plant, found some of the estimating assumptions were far off the mark. At storage tanks, for instance, the testing found releases of benzene were about 100 times higher than the estimated figures, although air concentrations were still within Alberta government safety levels. Under the industry's estimating procedure, the tanks were thought to be responsible for only about 12% of benzene emissions, although the tests found the figure was really 63%.

The purpose of preparing an accurate emissions inventory is not simply to meet a statutory reporting requirement. The objective of the Clean Air Act and

---

[1] http://www.theglobeandmail.com/servlet/Page/document/v5/content/subscribe?user_URL=http://www.theglobeandmail.com%2Fservlet%2Fstory%2FLAC.20080906.POLLUTANTS06%2FTPStory%2FEnvironment&ord=98941658&brand=theglobeandmail&force_login=true

state laws that are intended to implement the federal law at the local level is to protect the public. Regulators are supposed to rely on accurate industry reporting in order to assess the risks to the public from air pollution and to devise strategies to mitigate these risks. When refineries, gas-field operations, natural gas processing plants, and other industrial facilities under-report their emissions they are placing the public at risk, and are violating federal and state statutes.

In this chapter we provide a critical assessment of the current practices used by refineries and gas plants in developing emissions inventories. We believe these methods are flawed and lead to compounding errors resulting in the significant under-reporting of fugitive emissions. There are better methods and practices for developing accurate emissions inventories than are generally relied upon in the USA. This chapter provides guidance and recommendations on some of those best practices.

5.2   Methodology by which emissions inventories are prepared

The term 'emissions inventory' refers to the mass rate accounting of priority pollutants from the different sources within a manufacturing process. Both fugitive and point sources of emissions are required to be accounted for. These are not by any means total emissions, but only those emissions that are required to be reported. The reader should note that our discussions are restricted only to air emissions, but there are other regulated waste forms such as liquid, wastewater, and solid wastes that industry is required to report on.

In the USA emissions inventories at refineries, in gas processing plants and in oil and gas fields (as well as throughout the chemical industry) are prepared largely by means of applying *emission factors* to volume or mass production rates. In other words, the vast majority of reporting of air pollutants in the USA is by means of calculation and not actual monitoring using field instrumentation.

According to the US EPA:

> *An emission factor is a representative value that relates the quantity of a pollutant released to the atmosphere with an activity associated with the release of that pollutant. These factors are usually expressed as the weight of pollutant divided by a unit weight, volume, distance, or duration of the activity emitting the pollutant (e.g. kilograms of particulate emitted per megagram of coal burned). Such factors facilitate estimation of emissions from various sources of air pollution … In most cases, these factors are simply averages of all available data of acceptable quality, and are generally assumed to be representative of long-term averages for all facilities in the source category (i.e. a population average).*
>
> (http://www.epa.gov/air/aqmportal/management/
> emissions_inventory/emission_factor.htm)

The general US EPA equation for emission estimation is:

$$E = A \times EF \times (1 - ER/100),$$

where:

E = emissions
A = activity rate
EF = emission factor
ER = overall emission reduction efficiency (%).

In the USA information is embodied in multiple volumes of the *AP-42 Compilation of Air Pollutant Emission Factors*, which was first published by the US Public Health Service in 1968. In 1972, it was revised and issued as the second edition by the US EPA. General emission factors are available to the public. However, variations in the conditions at a given facility, such as the raw materials used, temperature of combustion, and emission controls, can significantly affect the emissions at an individual location. The EPA states in AP-42 and common sense mandates that emission factors are also a function of the age and condition of operating equipment.

The US EPA Protocol, dated November 1995, entitled *1995 Protocol for Equipment Leak Emission Estimates* (EPA-453/R-95-017, "the 1995 EPA Protocol") presents four different methods for estimating equipment leak emissions. The methods, in order of increasing refinement, are:

- Method 1: Average Emission Factor Method;
- Method 2: Screening Value Range Method;
- Method 3: Correlation Equation Method;
- Method 4: The Unit-specific Correlation Equation Method.

In general, a more refined method requires more data and provides more reliable fugitive hydrocarbon emission estimates. It is also more costly to implement and hence is not relied upon by many refineries. In the Average Emission Factor Method and the Screening Value Range Method, emission factors are combined with equipment counts to estimate emissions. This is the least-cost methodology. To estimate emissions with the Correlation Equation Method, OVA-measured concentrations (screening values) for all equipment components are individually entered into correlation equations or counted as either default zeros or pegged components.

In the Unit-specific Correlation Equation Method, screening and actual mass emissions are measured as discussed in Section VII of the CAPCOA guidelines discussed below for a select set of individual equipment components at a site and then used to develop unit-specific correlation equations and pegged source factors. Screening values for all components are then entered into these unit-specific correlation equations and pegged source factors to estimate emissions.

The US EPA's four different methods can be applied and used to estimate fugitive emissions. Detailed discussion of the methods is presented in the 1995

EPA Protocol. Another source is the American Petroleum Institute (API) document, dated July 1997, entitled *Calculation Workbook for Oil and Gas Production Equipment Fugitive Emissions*, which provides step-by-step example calculations using each of the estimation methods outlined in the 1995 EPA Protocol. However, some of the factors and correlation equations associated with the first three methods have been corrected and revised. Method 4 is not affected because it calls for the collection of site-specific data, which are then used to develop unit-specific correlation equations and factors. Component counting, component screening, and the leak quantification must use the methods specified in Sections V, VI, and VII of the guidelines for the unit-specific equations and factors to be acceptable to the local districts.

5.2.1 Method 1: Average Emission Factor Method

This method is recommended when no reliable screening data are available. However, we fail to see how any modern refinery or gas plant in the USA today cannot generate an accurate inventory of its emission sources and perform screening audits. There are no technological reasons for a facility not to be able to generate such information. We believe facilities that rely on this methodology are irresponsible.

In this method, average emission factors for refineries and marketing terminals are relied upon. The following five steps are used:

1. Components are separated into component types such as non-flange connectors, flanges, open-ended lines, pump seals, valves, and other components.
2. Each component type is further separated into service types such as gas, light liquid, or heavy liquid if there are different emission factors for each service type.
3. The total number of components in each group (component type/service type) is then determined.
4. The number of components in each group is multiplied by the corresponding average emission factor to obtain the subtotal of emissions from the group.
5. The subtotals of emissions from all groups are then added to provide the total emissions from the facility.

As an example CAPCOA (1999) assume there are 5000 components at a loading terminal. These components were inventoried into eight groups of component type/service type that correspond to the 1995 EPA Protocol average emission factors for marketing terminals. The number in each group is multiplied by the appropriate average emission factor in Table 5.1, and the total emissions estimate for the marketing terminal, 0.0944 kg/h, shown in Table 5.2.

The subtotals in Table 5.2 may be further multiplied by the number of operating hours in a year or quarter in order to estimate the mass emissions for the period.

To further illustrate the methodology we provide additional emission factors for uncontrolled emissions below. If a facility has control equipment, such as

Table 5.1 1995 EPA Protocol refinery average emission factors

Component type	Service type	Emission factor (kg/h/source)[a]
Valves gas	Gas	2.68E−02
	Light liquid	1.09E−02
	Heavy liquid	2.30E−04
Pump seals[b]	Light liquid	1.14E−01
	Heavy liquid	2.10E−02
Compressor seals	Gas	6.36E−01
Pressure-relief valves	Gas	1.60E−01
Connectors	All	2.50E−04
Open-ended lines	All	2.30E−03
Sampling connections	All	1.50E−02

[a]These factors are for non-methane organic compound emission rates.
[b]The light liquid pump seals factor can be used to estimate the leak rate from agitator seals.
Source: *1995 EPA Protocol for Equipment Leak Emission Estimates* (EPA-453/R-95-017, November 1995), which referenced the 1980 Petroleum Refining Study (EPA-600/2-80-075c, April 1980). These factors are based on the 1980 Petroleum Refining Study.

a condenser, the emissions can be multiplied by the control factor. The procedure is to calculate the control factor by subtracting the percentage control efficiency from 100 and then dividing that number by 100. For example, if the control efficiency is 90%, the control factor would be (100 − 90)/100 = 0.10. Control efficiencies may be listed on the equipment or in the equipment documentation. Alternatively, equipment suppliers can provide control efficiency values, but

Table 5.2 Sample calculation using the Average Emission Factor Method

Component type	Service type	Number of components	Average emission factor (kg/h/source)	Subtotals (kg/h)
Valves	Gas	28	1.3E−05	0.0004
	Light liquid	593	4.3E−05	0.0255
Pump seals	Gas	0	6.5E−05	0.0000
	Light liquid	25	5.4E−04	0.0135
Others	Gas	20	1.2E−04	0.0024
	Light liquid	85	1.3E−04	0.0111
Fittings	Gas	224	4.2E−05	0.0094
	Light liquid	4025	8.0E−06	0.0322
Total (kg/h)		5000		0.0944

these may require verification for older equipment that has been in service for many years.

Tables 5.3 and 5.4 provide some emission factors for different equipment. The emission factors are expressed in scientific notation, which means that the

Table 5.3 Emission factors for natural gas-fired engines[a]

Description	Pollutant	Emission factor	Control efficiency
Standard "rich-burn" engines			
May include:	CO	3.794E3 lb/MMCF natural gas[b]	Three-way catalyst
• Natural gas process heaters	NO_x	2.254E3 lb/MMCF natural gas[b]	
• Natural gas production, compressors	PM10	9.69E0 lb/MMCF natural gas[b]	CO – 80%[c]
• Natural gas production, flares excluding SO_2	PM2.5	9.69E0 lb/MMCF natural gas[b]	NO_x – 90%[c]
	SO_2	6.00E−1 lb/MMCF natural gas[b]	VOCs – 50%[c]
	VOCs	3.02E1 lb/MMCF natural gas[b]	
"Lean-burn" engines			
May include:	CO	5.68E2 lb/MMCF natural gas[b]	Oxidation catalyst
• Natural gas process heaters	NO_x	4.162E3 lb/MMCF natural gas[b]	
• Natural gas production, compressors	PM10	7.90E−2 lb/MMCF natural gas[b]	CO – 80%[c]
• Natural gas production, flares excluding SO_2	PM2.5	7.90E−2 lb/MMCF natural gas[b]	VOCs – 50%[c]
	SO_2	6.00E−1 lb/MMCF natural gas[b]	
	VOCs	1.204E2 lb/MMCF natural gas[b]	

[a]Report all "standard" engine emissions together, and report all "lean-burn" emission engines together. For facilities with both "standard" and "lean-burn" emission engines, report "standard" engines and "lean-burn" emission engines as separate emission units. Split the total fuel gas between the two different types of engines based on your best estimate of the relative amount of fuel burned in each type of engine at the facility. Group all natural gas combustion equipment with a standard "rich-burn" or lean-burn engines. For example, you may group all standard "rich-burn" engines, natural gas process heaters, production compressors, and flares together.
[b]The emission factors listed are derived from AP-42 Chapter 3.2 (Tables 3.2-2 and 3.2-3).
[c]The control factors listed above can only be used if documentation is on file showing that the catalyst was inspected and maintained. If actual control efficiencies are different than those listed above, use the actual control efficiency.
Source: obtained from the Michigan Department of Environmental Quality, Fact Sheet No. 9845 (Rev. 10/2006).

Table 5.4 Emission factors for various equipment

Description	Pollutant	Emission factor	Control efficiency
Process heaters[a]	CO	3.50E1 lb/MMCF natural gas	
	NO$_x$	1.40E2 lb/MMCF natural gas	
	PM10	3.00E0 lb/MMCF natural gas	
	SO$_x$	6.00E−1 lb/MMCF natural gas	
	VOCs	2.80E0 lb/MMCF natural gas	
Tank storage[b]			
Fixed roof tank – breathing loss	VOCs	3.6E1 lb/kgal-year crude oil (storage capacity)	Vapor recovery system – 95% Flare – 95%
Fixed roof tank – working loss	VOCs	1.1E0 lb/E3 gal crude oil (throughput)	Vapor recovery system – 95% Flare – 95%
Truck loading	VOCs	2.0E0 lb/E3 gal crude oil	Vapor recovery system – 95%
Gas dehydrators[c]			
Glycol dehydrator – Niagaran	VOCs	9.24E4 lb/year-GPM glycol[d]	Tube and shell condenser with flash tank – 90% Vapor recovery system – 95% Flare – 95%
Glycol dehydrator – Prairie du Chien	VOCs	1.94E4 lb/year-GPM glycol[d]	Tube and shell condenser with flash tank – 90% Vapor recovery system – 95% Flare – 95%
Glycol dehydrator – Antrim	VOCs	9.2E1 lb/year-GPM glycol[d]	Vapor recovery system – 95% Flare – 95%

Continued

Table 5.4 Emission factors for various equipment—cont'd

Description	Pollutant	Emission factor	Control efficiency
Amine plant			
	SO_2	3.76E3 lb/ton hydrogen SUL	
Fugitive emissions			
Light crude production	VOCs	1.44E1 lb/each-year valve	
Gas production	VOCs	3.6E0 lb/each-year valve	
Gas plant	VOCs	2.74E1 lb/each-year valve	

[a]Process heaters include process heaters as a separate emission unit if they were not grouped with natural gas-fired engines. The emission factors for process heaters come from the US EPA's Factor Information Retrieval (FIRE) data system, which can be accessed at http://cfpub.epa.gov/oarweb/index.cfm?action=fire.main. (Emission factors from Chapter 1.4 (Table 1.4-1) of the US EPA's AP-42 *Compilation of Air Pollutant Emission Factors* may also be used to calculate emissions from process heaters.)

[b]You may also use the US EPA Tanks 4.0 software to estimate emissions from tank storage. This software can be downloaded at www.epa.gov/ttn/chief/software/tanks/index.html.

[c]You may also use GRI-GLYCalc™ 4.0 software developed by the Gas Research Institute (GRI) to estimate emissions from glycol dehydrators. This software can be purchased at www.gastechnology.org.

[d]Year-GPM glycol = gallon per minute glycol circulated, averaged over 1 year.

Source: obtained from the Michigan Department of Environmental Quality, Fact Sheet No. 9845 (Rev. 10/2006).

decimal point has been moved. If the exponent is negative, move the decimal point to the left. If the exponent is positive, move the decimal point to the right. If the exponent is zero, the decimal point does not move. For example, if a number is expressed as 2.0E−1, move the decimal point one place to the left to get 0.20. If a number is expressed as 2.0E2, move the decimal point two places to the right to get 200. If a number is expressed as 2.0E0, the decimal point does not move. The number is 2.0.

The following are sample calculations:

1. For a Glycol dehydrator (Niagaran) equipped with a vapor recovery system, where 0.3 gallons per minute (GPM) of glycol is circulated, the VOC emissions would be calculated as follows:

$$0.3 \text{ GPM} \times 9.24\text{E } 4 \text{ lb/year} - \text{GPM} \times 0.0005 \text{ lb/ton}$$
$$\times (100 - 95)/100 = 0.69 \text{ tons VOC,}$$

where 0.3 GPM is the throughput, 9.24E4 lb/year-GPM is the emission factor, 0.0005 lb/ton is a conversion factor, and (100 − 95)/100 is the control factor.

2. For standard "rich-burn" engines with a properly maintained three-way catalyst where 4.25 MMCF of fuel gas was burned, the CO emissions would be calculated as follows:

$$4.25\text{MMCF} \times 3794 \text{ lb CO/MMCF} \times 0.0005 \text{ lb/ton}$$
$$\times (100 - 80)/100 = 1.61 \text{ tons CO,}$$

3. For lean-burn engines where 4.25 MMCF of fuel gas was burned, the CO emissions would be calculated as follows:

$$4.25 \text{ MMCF} \times 568 \text{ lb CO/MMCF} \times 0.0005 \text{ lb/ton} = 1.21 \text{ tons CO.}$$

The precision of the calculation procedures depends on the accuracy of the emissions source inventory as well as how representative the emission factors are of a particular facility. As shown by an earlier example (that of PXP), there are no assurances that companies are preparing and maintaining accurate inventories. Additionally, we must recognize that facility operators and owners are not required to determine whether or not the published average emission factors truly represent the operations of the facility. AP-42 notes that emission factors can be site specific and depend on the age and condition of equipment as well as the level of maintenance required to maintain them in operating conditions. As we note under Section 5.3.1, the assumption that all emission factors are equal or equally characterize any and all facilities is highly questionable in our opinion.

5.2.2   Method 2: Screening Value Range Method

The Screening Value Range Method was formerly known as the Leak/No Leak Method. This method uses the screening data (instrument screening values, "SVs") to calculate the mass emission rates based on the component leak level (below 10,000 ppm = no leak, or 10,000 ppm or greater = leak). Some parts of the country define leaks at lower levels than 10,000 ppmv. A region may choose to apply the 10,000 ppmv emission factors to all components above their leak definition. Such a policy will generally result in a conservative estimation of emissions. CAPCOA recommends that facilities that record individual screening values for each component may prefer to use the Correlation Equation Method (Method 3) or the Unit-specific Correlation Equation Method (Method 4).

The application of Method 2 (Screening Value Range) requires the completion of the following five steps:

1. Components are separated into component types (i.e. non-flange connectors, flanges, open-ended lines, pump seals, valves, others).
2. Each component type is further separated into service types (gas, light liquid, or heavy liquid) if there are different emission factors for the service types.
3. The total number of components in each group (component type/service type) with screening values below 10,000 ppmv is determined. The total number of components in each group with screening values of 10,000 ppmv or more is then determined.
4. The number of components in each subgroup (component type/service type/screening value range) is multiplied by the corresponding screening value range factor to obtain the subtotal of emissions from the subgroup.
5. The subtotals of emissions from all subgroups are added to give total emissions from the facility.

To illustrate the method we note the following example taken from the CAPCOA publication. The components on three large heaters at a refinery were screened and grouped by type, service (in this case all gas service), and screening value range. The number of components in each group was multiplied by the appropriate factor from Table 5.5 and the total emissions estimate for the heaters were found to be 2.56 kg/h as shown by the calculated values in Table 5.6.

The reader may refer to the CAPCOA publication for emission factors from other emission source types.

The facility owner and operator need to maintain an accurate inventory and should verify that emission factors are representative of the facility, the latter of which is not a requirement among the standards. Facilities do not have a standardized procedure for monitoring and classifying leaks and fugitive emissions, which introduces additional sources of errors into the inventory.

Table 5.5 1995 EPA Protocol refinery screening value range emission factors

Component type	Service type (kg/h/source)[a]	<10,000 ppmv	>10,000 ppmv (kg/h/source)[a]
Valves	Gas	6.0E−04	2.626E−01
	Light liquid	1.7E−03	8.52E−02
	Heavy liquid	2.3E−04	2.3E−04
Pump seals[b]	Light liquid	1.20E−02	4.37E−01
	Heavy liquid	1.35E−02	3.885E−01
Compressor seals	Gas	8.94E−02	1.608
Pressure-relief valves	Gas	4.47E−02	1.691
Connectors	All	6.0E−05	3.75E−02
Open-ended lines	All	1.5E−03	1.195E−02

[a]These factors are for non methane organic compound emission rates.
[b]The light liquid pump seals factor can be used to estimate the leak rate from agitator seals.
Source: *1995 EPA Protocol for Equipment Leak Emission Estimates* (EPA-453/R-95-017, November 1995), which referenced the 1982 Petroleum Refining Study (EPA-450/3-82-010, 1982). These factors are based on the 1980 and 1982 refining fugitive emissions studies.

5.2.3 Method 3: Correlation Equation Method

This method is based on the application of values presented in Table 5.7.

The following are recommended guidelines published in the CAPCOA guidance document:

- Default zero factors should apply only when the screening value, corrected for background, equals 0.0 ppmv (i.e. the screening value is indistinguishable from the background reading).
- The correlation equations apply for actual screening values, corrected for background and 9999 ppmv, and should be used for screening values up to 99,999 ppmv at the discretion of the local district.

Table 5.6 CAPCOA sample calculation using the Screening Value Range Method to estimate emissions from refinery heaters

Component type (service type)	Screening value range (ppmv)	Number of components screened	1995 Protocol screening value range factors (kg/h/source)	Subtotals (kg/h)
Valves (gas)	<10,000	164	6.0E−04	0.098
	>10,000	7	2.626E−01	1.838
Pressure-relief valves (gas)	<10,000	3	4.47E−02	0.134
	>10,000	0	1.691	0.000
Connectors (gas)	<10,000	703	6.0E−05	0.042
	>10,000	11	3.75E−02	0.413
Open-ended lines (gas)	<10,000	10	1.5E−03	0.015
	>10,000	2	1.195E−02	0.024

Table 5.7 1995 EPA correlation equations and factors for refineries and marketing terminals

Component type/ service type	Default zero factor (kg/h)[a]	Correlation equation (kg/h)[b]	Pegged factor (kg/h)[c]	
			10,000 ppmv	100,000 ppmv
Valves/all	7.8E−06	2.27E−06(SV)^0.747	0.064	0.138
Pump seals/all	1.9E−05	5.07E−05(SV)^0.622	0.089	0.610[d]
Others[e]/all	4.0E−06	8.69E−06(SV)^0.642	0.082	0.138
Connectors/all	7.5E−06	1.53E−06(SV)^0.736	0.030	0.034
Flanges/all	3.1E−07	4.53E−06(SV)^0.706	0.095	0.095
Open-ended lines/all	2.0E−06	1.90E−06(SV)^0.724	0.033	0.082

[a]The default zero factors apply only when the screening value (SV), corrected for background, equals 0.0 ppmv (i.e. the screening value is indistinguishable from background reading). The default zero factors were based on the combined 1993 refinery and marketing terminal data only; default zero data were not collected from oil and gas production facilities.

[b]The correlation equations apply for actual screening values, corrected for background, between background and 9999 ppmv, and can be used for screening values up to 99,999 ppmv at the discretion of the local district.

[c]The 10,000 ppmv pegged factors apply for screening values, corrected for background, equal to or greater than 10,000 ppmv, and are used when the correlation equations are used for screening values between background and 9999 ppmv. The 100,000 ppmv pegged factors apply for screening values reported pegged at 100,000 ppmv and are used when the local district authorizes use of the correlation equations for screening values between background and 99,999 ppmv.

[d]Only three data points were available for the pump seals 100,000 ppmv pegged factor.

[e]The "other" component type includes instruments, loading arms, pressure-relief valves, vents, compressors, dump lever arms, diaphragms, drains, hatches, meters, and polished rod stuffing boxes. This "other" component type should be applied for any component type other than connectors, flanges, open-ended lines, pumps, or valves. However, if an acceptable emission estimate exists that more accurately predicts emissions from the source, then that emission estimate applies (e.g. positive-flowing junction boxes in SCAQMD).

Source: SBCAPCD Report, dated 1 May 1997, entitled *Review of the 1995 Protocol: The Correlation Equation Approach to Quantifying Fugitive Hydrocarbon Emissions at Petroleum Industry Facilities.* Technical corrections and adjustments were made to the refineries and marketing terminals bagged data, obtained by use of the blow-through method, to account for the hydrocarbon leak flow rate.

- The 10,000 ppmv pegged factors apply for screening values, corrected for background, equal to or greater than 10,000 ppmv and should be used when the correlation equations are used for screening values between background and 9999 ppmv. The 100,000 ppmv values are used when the local district authorities use the correlation equations for screening values between background and 99,999 ppmv.
- Where multiple instrument screening values apply as with quarterly inspections, the average of the calculated mass emission estimates during the reporting period should be used.

The following steps are applied in this method:

1. Record each individual component screening value.
2. Group the data component type into three categories of screening ranges: default zeros range, correlation equations range, and pegged source range.
3. Multiply the number of components with instrument screening values in the default zeros range by the appropriate default zero factors.

4. Enter each individual component screening value that is within the correlation equations range into its appropriate correlation equation.
5. Multiply the number of components with instrument screening values in the pegged range by the appropriate pegged values.
6. Sum up all the calculated emissions to obtain an estimate of the total emissions from the facility.

The CAPCOA guidance document provides a detailed example that the reader may refer to. This method is believed to be more accurate than the previous two methods but is cumbersome to apply.

5.2.4   Method 4: The Unit-specific Correlation Equation Method

This method requires the facility operator or owner to collect site-specific data, which are then used to develop unit-specific correlation equations and factors. These are subject to approval by state regulatory agencies. In principle we believe this to be the most precise method because the facility in essence develops a site-specific emission factor or correlation for emission factors that is specific to the equipment and controls.

This type of empirical approach is reasonable and is analogous to the development of control equations, which are generally used throughout the chemical industry in guiding operators in the control of chemical reactors. Generally the development of such empirical correlations that can then be applied to standard emission calculations that are well documented for different types of components like valves, diaphragms, seals, flanges, and other components implies further that the empirical correlations will have well-defined statistics such as mean, standard deviation, range, and confidence limits.

Examples of published empirical correlations include the various API (1987, 1991, 1996, 1997) algorithms for determining evaporation losses from storage tanks and product loading/unloading terminals, and leak-rate correlations for converting leak screening data to emissions rates (GRI Canada, 1998; US EPA, 1995b). But these published correlations carry with them the same inherent flaws that accompany emission factors themselves; in other words they are based on average reported values and are not necessarily representative of a facility. We are still faced with applying an underlying assumption that on average the calculated emissions are the same for any one facility and only production rates influence the amount of emissions.

In all of these methods and indeed the basis for applying AP-42 or any published emission factor or correlation to calculating air emissions is that, on average, all facilities will generate the same amount of pollution per unit of production since the manufacturing and control technologies are approximately the same for all facilities. Because the industry appears to rely almost entirely on calculation methods that use average emission factors that are not verified as being site specific, we in part have concluded that there is gross under-reporting of emissions. More concrete evidence to support this claim is presented later in the chapter.

Facility owners and operators argue that the industry on the whole tends to overestimate air emissions because of so-called conservative assumptions applied to the calculations. For example, in the PXP litigation, the company argued that although its quarterly reporting of emissions from piping aperture in its gas plant and field gathering operations was not based on an accurate inventory of flanges versus connectors, it had pointed out that the total number of components was accurate. It assumed for calculation purposes that all the emissions came from vapor service and ignored any liquid service. According to this argument, since the emission factor for flanges is nearly twice that of connectors, a conservative assumption that all the service is from vapor handling operations should more than compensate for the lack of an accurate inventory of flanges versus connectors. PXP has argued that this would make the emissions calculations conservative and the reported emissions should be greater than actual emissions; however, this is not a reasonable argument if one cannot at least provide comparative calculations to substantiate the magnitude of the conservatism. Furthermore, we could not find a single example where a US company has attempted to establish a confidence limit on its emissions reporting or performed at the very least a reality check by obtaining actual measurements and then making comparisons to its calculated facility emissions for any priority pollutant.

5.3   Inherent flaws that contribute to biased reporting

In the previous section we explained the application of the generally accepted methodology and calculation procedures applied to quantifying the yearly average emissions from a facility. Assuming the methodology is rigorously applied, then why are there criticisms from not just us, but many credible and authoritative independent organizations that argue that US refineries generally under-report their air emissions? Either facilities are not accounting for all of their emission sources or there are fundamental flaws in the methodology and calculation methods that result in a low bias of calculated VOC emissions.

We have analyzed this question and have concluded that there are a number of reasons that account for a low bias in emissions reporting. The following are our observations based on a review of the literature, our combined industry experience that extends more than a half a century, and intimate involvement in a number of environmental litigations.

5.3.1   Assumption that all emission factors are equal

The basis for the application of any industry-published emission factor is the assumption that all facilities on average will generate about the same amount of pollution per unit of production if the same technologies and controls are used. Further to this assumption, most facilities and the API argue that since published emission factors are averages, calculated emissions should be viewed as yearly or rather long-term averages. In other words, although there may be excursions in releases, if one were to take measurements of actual emissions over a sufficiently long period of time, the average of the measurements would be in agreement with calculated mass emissions.

Published emission factors for the most part are based on industry-reported averages. We must assume that a sufficiently large enough population of facilities has been sampled to obtain representative emission factors such that variations among equipment performance are included in the published factor. To our knowledge this has never been substantiated. While AP-42 provides a semi-qualitative rating of the accuracy of published emission factors, it publishes neither a range nor a standard deviation for emission factors. It only publishes a single value and advises the user that there is a higher level of confidence in some values versus others. But again we emphasize that published emission factors are based on values obtained and reported by the industry itself. We are not implying that these values are misrepresentations, but rather that for the most part when such values were originally obtained from measurements they were based on normal, steady-state operation of equipment. This is a very important observation that we discuss at greater length later in the chapter.

The EPA has acknowledged in discussions on its website and in AP-42 itself that emission factors can be site specific. Not only are there variations among the same technologies used, especially controls, but the age and condition of equipment as well as site-specific practices can dramatically impact the value of any emission factor. Logically, let us consider a 35-year-old electrostatic precipitator and one that is less than 5 years old. It is common sense the newer equipment performs better, has a higher service factor, and is more efficient.

The most serious shortcoming in relying on any published emission factor is that published values are based on measurements obtained during steady-state, continuous operations. Published emission factors do not capture upsets, breakdowns, excursions, or other operational conditions that can lead to episodic releases. Older equipment and operations are prone to these events and certainly more frequently than newer controls and process equipment that come with warranties.

Another consideration is that not all facilities apply the same protocols to performing maintenance. Some facilities do scheduled inspections, which means that the facility identifies leaks and emissions at the time of the inspection but may not understand when some of these episodes started or how long they have gone on before corrective actions were taken. Certainly older facilities with aging infrastructure that are not well maintained or have long lead times between performing inspections and maintenance on equipment present a high degree of uncertainty and there is simply no basis to argue that published emission factors are representative of the actual site conditions even on average.

The age and condition of equipment, especially air pollution controls, unquestionably have an impact on the accuracy of emission factors that are used for calculating air pollution levels for a facility. The World Bank Organization (WBO), as an example, reports ranges of emissions factors and makes the

Table 5.8 World Bank Organization reported emission factors (lb/Bbl crude)

Pollutant	Average	Low	Upper
Particulate matter	2.778E−01	3.473E−02	1.042E+00
Sulfur oxides	4.515E−01	6.946E−02	2.084E+00
Sulfur oxides with sulfur recovery	3.473E−02		
Nitrogen oxides	1.042E−01	2.084E−02	1.736E−01
VOCs	3.473E−01	1.736E−01	2.084E+00
BTX	8.682E−04	2.605E−04	2.084E−03

Source: *Pollution Prevention and Abatement Handbook*, World Bank Group, Washington, DC, July 1998.

distinction that the higher values reported are more typical of older facilities with aging infrastructure (see Table 5.8). Relying on default values that are reported by state regulatory agencies or AP-42 without any independent verification as to whether or not those values truly characterize the average conditions for a facility is not a reasonable approach. In examining the WBO's published values we see that there are several orders of magnitude differences between the high and low values for some reported emissions. Relying on an average value as in the case of the AP-42 procedure does not support a general industry argument that calculated emissions are conservative and overstate the pollution from a facility.

5.3.2 Omissions and mischaracterizations

The introduction to this chapter provided but a few examples of practices and companies that do not provide accurate accounting of their emissions. In the case of PXP, while the company argues that it applies a conservative basis to calculating fugitive emissions by assuming that all of its service is with gas handling as opposed to liquid handling, it simply ignored quantifying, documenting, and reporting thousands of incidents that are leaks and fugitive emissions. For all practical purposes it handed off its reporting obligations to a third party and failed to share with the consultant the fact that emissions and leaks resulting in fugitive emissions were occurring on an ongoing basis. This is not a simple omission in terms of meeting a reporting requirement. The fact is that PXP did not provide its maintenance personnel with monitoring instruments and hence never had an accurate accounting of its emissions. But its failure to even share the information on the numbers of emission incidents from its maintenance records with the consultant responsible for filing quarterly emissions reports means that both the public and regulators were denied access to critical information that could have been applied towards making knowledgeable decisions with regard to public health risks.

The PXP example is by no means an isolated example. The Waxman report to Congress clearly demonstrates that the industry does not rely on standardized procedures that accurately identify and quantify fugitive emissions. A fivefold difference accounting for the emissions from leaking valves means that there is an equivalent magnitude of fugitive emissions not accounted for. This leaves the public and regulatory agencies with a gross misrepresentation of the potential risks to inhalation exposures within communities.

A memorandum issued by Shine (2007) of the US EPA documents numerous incidents throughout the industry where omissions and misrepresentations of fugitive emissions are ongoing. The following summarizes the various incidents that are noted as typical omissions in reporting:

- exclusion of upsets, malfunctions, startups, and shutdowns from emissions inventories;
- omission of sources that are unexpected or not measured, such as leaks in heat exchanger systems or emissions from process sewers;
- exclusion of emission events such as tank roof landings;
- improper characterization of input parameters for emission models such as not using actual tank or material properties in the AP-42 tank emission estimation methodologies.

Shine's memorandum points out that the current US National Emissions Inventory does not identify upsets, startups, or shutdowns as emission events, nor is the data specifically requested from the reporters (the states). To understand the order of magnitude of these omissions and upsets in relation to routine operations, the EPA reviewed the emission inventory data from the Texas Commission on Environmental Quality (TCEQ) for the 2004 reporting year. This dataset contains emissions data for 30 of the approximately 150 US refineries and accounts for over 25% of the US refining capacity. Additionally, the TCEQ inventory identifies emissions from routine events separately from upsets, startups, and shutdowns, so a comparison of reported emissions is possible.

Shine reported that, in general, the quantity of emissions reported as non-routine is smaller than that of the routine emissions. For VOC-unclassified contaminant, emissions of upsets and various malfunctions, startups, and shutdowns were 5% of the emissions reported from routine events (578 TPY versus 11,032 TPY). However, for some compounds, such as 1,3-butadiene, emissions from these incidents accounted for as much as 20% of the routine emissions (19.8 TPY versus 91 TPY). The investigator goes on to note that, for certain types of emission points, emissions from startups, shutdowns, and malfunctions comprise the majority of the emissions.

Shine has further noted that the comparison was done between *reported* upsets and the unstable or transient events and *reported* routine emissions. This comparison does not consider events such as upsets and shutdowns/startups/malfunction events, which are not properly characterized and reported to begin with.

The EPA memorandum notes that there are emission events that are not measured and further that there are many events that are not even characterized or reported in inventories. For example, monitoring of cooling tower water return for VOC is required at some refineries because of state permitting and RACT rules, but these are not required for refineries at the Federal level (e.g. by the Petroleum Refinery MACT standard).

Speciation of the VOC to individual hazardous air pollutant (HAP) compounds is typically not required. Because there is no requirement to monitor for leaks, there is, in effect, no systematic mechanism for facilities to identify, quantify, and control emissions in a timely way.

Further, there is potential for high emissions of VOCs and HAPs from such events. For example, Shine has noted that in one release report submitted to the National Response Center in 2006, a facility initially reported potential emissions of 700 lb/day each of benzene, toluene, and xylene from a reformer unit cooling tower, based on sampling of their cooling water return and the expected composition of the process streams that were being cooled. Upon further analysis and speciation of the cooling water, however, the facility submitted a final report indicating that the exchanger had leaked 800 lb/day of propane and isobutane for approximately 8 days. The EPA's memorandum noted that the subject facility monitored the tower and this is the reason why the leak was identified and reported. However, many refineries do not conduct routine cooling tower water monitoring.

In a sampling of the refining industry to be used to supplement the EPA's emissions inventory for the purpose of risk modeling, the EPA surveyed 22 refineries and requested emissions of benzene. Of the 22 facilities surveyed, only three indicated that they had sampled their cooling towers for leaks. The remaining facilities that did report emissions used AP-42 VOC emission factors for cooling towers and an assumed speciation for benzene. Five facilities simply reported no emissions at all.

Shine further noted that emissions from the delayed coking process are overlooked. In the measurements conducted at an Alberta refinery, the coker area was found to contribute to over 15% of the site VOC emissions and 26% of the benzene emissions. The measurements were made when the coke from the delayed coking unit was being drilled (after full water quench) and when it was not. Emissions were high when the coke was being drilled. Shine stated in her memorandum that, currently, US refiners do not report any fugitive emissions of VOCs or benzene from the delayed coking cutting/drilling/coke recovery process.

Additional omissions were identified by Shine regarding the wastewater treatment emission estimates provided by US refineries. A recent Bay Area (BA) AQMD study evaluated collection system emissions for five Bay Area refineries. Utilizing extensive sampling, flow measurements, and detailed TOXCHEM+ modeling, the study showed that four of the five refineries underestimated the VOC emissions from their wastewater collection system. Two refinery estimates were within a factor of 2 of the BA AQMD estimate (one higher and one lower), but one refinery had underestimated its emissions by a factor of 40 and another refinery underestimated its emissions by a factor of 1400. In reviewing the emission estimates reported by the residual risk survey respondents for wastewater collection and treatment systems, Shine also noted low estimates for several refineries.

Shine's memorandum also notes an omission observed by one of the present authors at several refineries visited over the past 30 years, namely that of floating-roof landings and collapses. Floating roofs are an effective method of controlling VOC emissions from storage tanks because they prevent direct contact of the stored liquid with ambient air and limit the formation of a saturated vapor in the headspace of the tank. However, if the liquid level in the tank is lowered to below the surface of the floating-roof support legs, the roof will land on its legs, creating a saturated vapor space and limiting the control efficiency of the floating roof. The TCEQ has estimated that under-reported landing loss emissions in the Houston–Galveston area alone totaled over 7000 tons of VOCs in 2003 according to Shine. The EPA recently updated AP-42 to include API methodology for calculating roof landing losses. As with cooling tower leak monitoring, there does not appear to be a systematic mechanism at the Federal level for facility owners to identify, quantify, and control these events, although the TCEQ has proposed rulemaking to limit the circumstances under which tank landings occur and has issued guidelines for reporting of these events in their inventories.

To further illustrate the under-reporting from tank roof failures, we examined a historical event from the Gulf Hooven refinery in Ohio (*Baker et al.* v. *Chevron USA*). In this example we found an 81,500 barrel (Bbl) gasoline tank in which the roof had sunk. The refinery had characterized this event as a tank landing and reported that it calculated the fugitive emissions over a 7-day event to be around 3500 lb of VOC fugitive emissions based on applying the assumption that the tank behaved as a fixed roof installation during the failure and thus an AP-42 emission factor applied. However, a close examination of the documented event showed that the roof had actually sunk and that there was free product on the surface of the roof for a period of a week (Bates Stamp Document AIR000043). The following is our series of calculations for the fugitive emissions:

1. Using AP-42, Chapter 7 for Storage Tanks:
   a. From Table 7.1-7 (AP-42), the average ambient temperature for the month of August (when the event occurred) for the nearest city (Columbus) = 61.7°F (289.65 K).
   b. From Table 7.1-9(AP-42), average annual wind speed = 8.5 mph = 3.8 m/s.
   c. From Table 7.1-2 (AP-42), the true vapor pressure for a range of gasolines:
      i. RVP7: 3.5 psi @ 60°F
      ii. RVP15: 8.16 psi @ 60°F
2. From a plot plan of the refinery, we found the elevation of the tank in question (492.5 ft). Since the elevation of the tank and the holding volumes were known, the area of the exposed gasoline on the surface of the sunken roof was calculated as follows:
   $A = 81{,}500$ Bbl $\times$ 42 gal/Bbl (7.48 gal per cubic ft $\times$ 492.5 ft) = 929.2 ft^2 or 86.4 m^2.

3. Using EPA's calculation procedure from "Risk Management Guidance for Offsite Consequence Analysis" (US EPA Publication EPA-550-B-99-009, April 1999), the fugitive emissions from a pool of liquid may be estimated from the following formula:

$$E = \{0.1268 A P M^{0.667} u^{0.78}\}/T,$$

where

E = evaporation rate of the liquid (kg/min)
u = wind speed just above the pool liquid surface (m/s)
M = pool liquid molecular weight (from Table 7.1-2 in AP-42); for RVP7 $M = 68$, for RVP15 $M = 60$
A = pool surface area (m²)
P = vapor pressure of the pool liquid at the pool temperature (kPa); from Table 7.1-2 (AP-42) $P = 3.5$ psi $= 24.13$ kPa for RVP7, $P = 8.162$ psi $= 56.3$ kPa for RVP15
T = pool liquid ambient temperature (K; assumed to be same as ambient conditions).

The calculation for evaporation of RVP7 gasoline is:

$$E = 0.1268(86.4 \text{ m}^2)(24.13 \text{kPa})(68)^{0.667}(3.8 \text{ m/s})^{0.78}/289.6 \text{ K}$$

$$= 41.4 \text{ kg/min}.$$

Since the evaporation took place over a 7-day period, the total mass of gasoline that evaporated was:

$$41.4 \text{ kg/min} \times 60 \text{ min/h} \times 24 \text{ h/day} \times 7 \text{ days}/0.454 \text{ kg/lb} = 957,880 \text{lb}.$$

Assuming 1.2% (mass basis) of the gasoline contained benzene, the amount of benzene released as fugitive emissions was $0.012 \times 920{,}073$ lb $= 11{,}494$ lb.
Repeating the calculation for RVP15:

$$E = 0.1268(86.4)(56.3)(60)^{0.667}(3.8)^{0.78}/289.6 = 92.6 \text{ kg/min}.$$

The evaporation over 7 days was:

$$92.6 \times 60 \times 24 \times 7/0.454 = 2{,}055{,}917 \text{ lb}.$$

Assuming 1.2% benzene in the gasoline:

$$0.012 \times 2{,}055{,}917 = 24{,}671 \text{ lb benzene}.$$

These are approximations of course, but the calculations illustrate quite clearly how a mischaracterization of events can lead to significant misreporting of emissions. There is a multiple of 300–645 times more VOC emissions over those reported in the 1985 event. Furthermore, the refinery never made any mention of benzene emissions at the time of reporting the incident or any time thereafter.

One of the more disturbing observations reported by Shine is that in an Alberta refinery measurement study, emissions of VOCs were 30 times higher and emissions of benzene were 100 times higher than emissions calculated using AP-42 equations. The reason for this under-reporting is that the AP-42 equations require a number of inputs about the tank and material characteristics and storage conditions. Mischaracterization of these inputs leads to erroneous results. The API points out that when actual measurements indicate unexpectedly high emissions, environmental conditions may be outside the scope of the method. But then how is the refinery to know when and when not to apply the calculation method properly? These concerns are sources of uncertainty that can explain differences of the order of 2 or 3, but they do not explain differences that are of the order of 30–100. Given the magnitude of the difference, either emissions are zero most of the time (when events are not on the high side) or the annual emissions estimates are grossly understated.

Further to this, Shine has noted that there are numerous examples of tank maintenance issues that, if not characterized properly, would lead to erroneous results. One example cited is: on 11 March 2003, the South Coast Air Quality Management District (SCAQMD) filed suit against BP West Coast Products, LLC. Most of the allegations accuse the company of failing to properly inspect and maintain 26 storage tanks equipped with floating roofs, as required under SCAQMD Rule 463. SCAQMD inspections revealed that more than 80% of the

Figure 5.1 Steam losses from traps at a refinery in Jordan.

tanks had numerous leaks, gaps, torn seals, and other defects that caused excess emissions (Whetzel, 2003).

In addition to the areas noted in the Shine memorandum, we take issue with those facilities that rely on contaminated groundwater for process operations. The use of contaminated groundwater to meet heat exchange and other non-contact process water requirements introduces additional sources of emissions that are not accounted for and not reported in inventories. Consider that there can be literally several thousand steam traps in use at a typical refinery. These generally fall way down on the pecking list for priority maintenance. Refineries that rely on groundwater that has been contaminated with gasoline constituents from leaking tanks, or from many years of poor housekeeping practices, introduce fugitive emissions that will contain benzene, VOCs, and other gasoline components.

The Jordan refinery in the Amman/Zarqa region is an example of a facility one of the authors spent a fair amount of time at. The photographs shown in Figures 5.1–5.3 show a few examples of many hundreds of steam ejections and losses that were documented for this facility. Figure 5.4 shows waste oil spillage, Figure 5.5 shows leaking valves, Figure 5.6 shows a landfill area where discarded drums of tetraethyl lead were deposited, and Figures 5.7 and 5.8 show wide site contamination of crude from filling line

Figure 5.2 Other examples of steam losses.

Figure 5.3 Additional examples of steam losses.

Figure 5.4 Waste oil spills.

Figure 5.5 Leaking valves.

Figure 5.6 A landfill area where discarded drums of tetraethyl lead were deposited.

Figure 5.7 Crude spillage from tanker line drainage. The facility was not using hoses with caps so when hoses were placed on the ground, drainage occurred.

Figure 5.8 Site contamination from poor housekeeping practices.

drainage and generally poor housekeeping. It is certainly not a leap in logic to assume that the groundwater aquifer beneath this facility has more likely than not been impacted by crude and waste oil, gasoline components including benzene and lead. Since the author was a guest at the facility with a restricted assignment of identifying pollution prevention opportunities for water savings only, the management was not receptive to the idea of either sampling the groundwater or obtaining some measurements of the fugitive emissions around steam traps. Hence, we can only note the physical conditions observed and allow our conclusions to be criticized as unlikely leaps in logic, which some US oil companies have argued when similar observations were made at other facilities.

Logically, why would one not expect that steam traps themselves are a significant source of fugitive hydrocarbon emissions for a refinery that relies on groundwater that is contaminated? If we follow the AP-42 methodology for accounting for fugitive emissions, any VOCs and benzene resulting from steam losses would never be accounted for because a facility would argue that these apertures are used for water condensate collection of non-contact steam, and are thus not part of the hydrocarbon fugitive emissions inventory. After all, steam traps are intended to capture steam and condense the water for recycling, and hence are not normally sources of emissions; but let's face it, this argument has validity only if there is a 100% effective leak detection and repair (LDAR) program and contaminated groundwater is not the source of non-contact process water.

The underestimation of VOC emissions by the industry is not a revelation by any means. The Texas Air Quality Study-2000 noted that ambient concentrations of highly reactive VOCs were found to be 10–1000 times higher than were reported in the Texas emission inventory for that year, and the NARSTO Emission Inventory Assessment necessitated that reported VOC emissions be multiplied sixfold before models and ambient measurements correlated. The EPA's Office of Inspector General, in a 22 March 2006 report, specifically recognized the problem of under-reporting of VOC emissions from the refining sector and concurred with the Agency shifting towards more direct, continuous monitoring and measurement of emissions from all major sources.

The above provides sufficient cause for concern that the industry on the whole is likely not only sidestepping its reporting requirements, but misrepresents its emissions of dangerous toxins. Communities living near refinery operations are in all likelihood misinformed about the air quality they breathe and are unwittingly presented with clear and present dangers that the industry has chosen to ignore for decades.

5.4 Toxic Release Inventory

The Toxic Release Inventory was established in response to the 1984 methyl isocyanate release that killed thousands of people in Bhopal, India. Shortly after that event there was a serious chemical release at a sister plant in West Virginia. These incidents underscored demands by industrial workers and communities in the USA for information on hazardous materials. Public interest and environmental organizations accelerated demands for information on toxic chemicals being released "beyond the fence line" – outside of the facility. Against this background, the Emergency Planning and Community Right-to-know Act (EPCRA) was enacted in 1986.

The EPCRA's primary purpose is to inform communities and citizens of chemical hazards in their areas. Sections 311 and 312 of the EPCRA require businesses to report the locations and quantities of chemicals stored on-site to state and local governments in order to help communities prepare to respond to chemical spills and similar emergencies. EPCRA Section 313 requires the EPA and the States to annually collect data on releases and transfers of certain toxic chemicals from industrial facilities, and make the data available to the public in the Toxic Release Inventory (TRI). In 1990 Congress passed the Pollution Prevention Act, which required that additional data on waste management and source reduction activities be reported under the TRI program. As stated on the EPA's website, "The goal of TRI is to empower citizens, through information, to hold companies and local governments accountable in terms of how toxic chemicals are managed." The EPA compiles the TRI data each year and makes it available through several data access tools, including the TRI Explorer (www.epa.gov/trieaplorer/) and Envirofacts (www.epa.gov/enviro/). There are other organizations that also make the data available to the public through their own data access tools, including Unison Institute, which puts out a tool called "RTKNet", and Environmental Defense, which has developed a tool called "Scorecard".

The TRI program actually began in 1987. The EPA has issued rules to roughly double the number of chemicals included in the TRI to approximately 650. Seven new industry sectors have been added to expand coverage significantly beyond the original covered industries, i.e. manufacturing industries. Most recently, the Agency has reduced the reporting thresholds for certain persistent, bioaccumulative, and toxic (PBT) chemicals.

Over the past 8 years the former Bush Administration attempted to dismantle this program and/or bar public access to reporting information under the TRI. Frankly, we are not quite sure why the administration attempted to do so because there is substantial evidence to support that the inventory is misleading and greatly under-reports emissions to begin with.

The industry sector notebook published by the US EPA (1995a) noted that at that point in time the petroleum refining industry was releasing 75% of its total TRI poundage to air, 24% to water (including 20% to underground injection and 4% to surface waters), and 1% to land. This profile has not changed much over the last decade. In comparison to other TRI industries, average profiles are 59% to air, 30% to water, and 10% to land. This observation characterizes the petroleum industry as the most polluting in terms of air emissions compared to all other industry sectors.

The Waxman congressional study, the EPA Shine memorandum, the EIG study, and the examples described in the previous section all raise suspicions that the TRI is not a reliable source for air emissions reporting by the industry. In short, there is no basis to rely on the TRI as a basis for determining air quality and for health risks assessments.

De Marchi and Hamilton (2006) have performed a critical assessment of the TRI. Among other conclusions these investigators note that self-reported regulatory data are hard to verify. They compared air emissions reported by plants in the Toxic Release Inventory with chemical concentration levels measured by EPA pollution monitors. They find that the large drops in air emissions reported by firms in the TRI are not always matched by similar reductions in measured concentrations from EPA monitors.

For the 2000 reporting year, 23,484 facilities submitted 91,513 reports about the emissions and transfers of the approximately 650 chemicals tracked in the TRI. Between 1988 and 2000, the data reported to the EPA indicated that releases of the core TRI chemicals by manufacturing plants dropped by 48%, with part of the reduction attributed to scrutiny generated by public release of the data. Persistent questions about the operation of the TRI program, however, are generated because the data are self-reported.

Self-reporting is a clear conflict of interest. Konar and Cohen (1997) established that firms with the largest adverse market reactions subsequently reduced their TRI releases more. Khanna et al. (1998) found that negative stock market reactions to TRI information lead to a reduction in the release of toxics on-site but increase the waste shipped off-site. Khanna and Damon (1999) and Arora and Cason (1996) trace how firm-level benefits and costs influenced company choices about whether to join the EPA's voluntary 33/50 program, which targeted reductions in 17 TRI chemicals. All of this means that there is a direct financial incentive to report reductions in pollution on the TRI. While that is all to the good, if there is no independent verification of the reductions, why should the public have any more faith in this self-reporting program than it has in the home mortgage institutions that have failed by misrepresentation and fraudulent practices?

The text of the law that created the TRI, the Emergency Planning and Community Right-to-know Act of 1986 (EPCRA), explicitly states that facilities need not engage in substantial effort to derive their TRI figures. In order to provide the information required under this section, the owner or operator of a facility may use readily available data (including monitoring data) collected pursuant to other provisions of law or, where such data are not readily available, reasonable estimates of the amounts involved. Nothing in this section requires the monitoring or measurement of the quantities, concentration, or frequency of any toxic chemical released into the environment beyond that monitoring and measurement required under other provisions of the regulation (US Senate Committee on Environment and Public Works, 1990). The legislation does not set a particular standard for accuracy and gives wide latitude for calculated emissions.

The US EPA (1990) estimated that nearly one-third of the facilities that were required to report toxic emissions under the TRI program did not file reports in its first years of operation (i.e. 1987 or 1988 emissions). In recent years environmental groups have complained about the potential for firms to overestimate their reductions in emissions and have labeled some declines in reported TRI figures as "phantom" or "paper" reductions. For example, the National Wildlife Federation (Poje and Horowitz, 1990) surveyed 29 facilities about their toxic releases and transfers for 1987 and 1988 and determined that the largest reported decreases came from "changes in reporting requirements, analytical methods, and production volume, and not from source reduction, recycling, or pollution abatement".

De Marchi and Hamilton's study cites various EPA studies in which plants were surveyed about changes in their reported TRI figures, finding that at least half of the net change in TRI figures came from a mixture of real and paper changes. The investigators cite EPA studies in which site surveys were used to verify the accuracy of TRI reports that consistently found that some plants had difficulties in determining whether their use of a chemical surpassed thresholds for reporting and that plants were more likely for air releases to have monitoring data on stack releases than on fugitive emissions (e.g. leaks).

5.5 IPCC assessment and best international practices

The Intergovernmental Panel on Climate Change (IPCC) was established by the World Meteorological Organization (WMO) and the United Nations Environment Programme (UNEP) in 1988. Its main objective was to assess scientific, technical, and socio-economic information relevant to the understanding of human-induced climate change, potential impacts of climate change and options for mitigation and adaptation.

The IPCC Guidelines have a three-tier approach for estimating fugitive emissions from oil and gas activities:

- Tier 1: Top-down average emission factor approach
- Tier 2: Mass balance approach
- Tier 3: Rigorous bottom-up approach.

Tier 1 is essentially Method 1 described earlier. The IPCC notes this to be the simplest as well as the least reliable approach. It is a top-down approach in which average production-based emission factors are applied to reported oil and gas production volumes. The IPCC states that this method is, at best, an order-of-magnitude approach, and should only be used as a last resort.

Tier 2 is a mass balance approach. It is primarily intended for application to oil systems where the majority of the associated and solution gas production is vented or flared. In these cases, the total amount of associated and solution gas produced with the oil is assessed, and then control factors are applied to the results to account for conserved, reinjected, and utilized volumes. The result is the amount of gas either flared or lost directly to the environment (whether through equipment leaks, evaporation losses, or process venting). The flared, utilized, and conserved volumes are determined from available production accounting data and engineering estimates. The rest of the gas, by difference, is lost directly to the atmosphere. The reliability of this approach increases as the portion of the total gas conserved, utilized, or reinjected decreases. The total amount of associated gas per unit volume of oil production is given by the gas-to-oil ratio within the target oil fields. The amount of solution gas or product volatilization per unit oil production is determined from the change in product vapor pressure between the inlet separator at the field production facility (i.e. the vessel operating pressure) and the refinery inlet (e.g. a Reid vapor pressure of 30–55 kPa). There is not much information on the accuracy of this approach; however, in principle it is more reliable than the use of average emission factors.

Tier 3 relies on the rigorous assessment of emissions from individual sources using a bottom-up approach, and requires both process infrastructure data and detailed production accounting data. It may also include actual measurement work. The results are then aggregated to determine the total emissions. The

IPCC Guidelines do not establish criteria for conducting the individual source assessments. Rather, they refer to several recently published emission inventories for the oil and gas sector that are deemed to be representative of a rigorous bottom-up approach. Consequently, there is a wide range in what potentially may be classified as a Tier 3 approach, and correspondingly in the amount of uncertainty in the results.

Regardless of the methodology applied, there are a limited number of tools that are available. These tools are:

- emission factors;
- computer models and simulators;
- direct measurement techniques;
- indirect measurement techniques;
- activity data;
- production statistics;
- data infrastructure.

5.5.1 Emission factors

We have already discussed the application of emission factors. Average emission factors have been developed and published by environmental agencies and industry associations. The proper way to view these is that they are statistical values that may be expected to provide reasonable results when applied to a large population of applicable sources (e.g. for regional and national emission inventories), but they are not reliable when applied to individual or small numbers of sources, unless they have been verified as being representative of a particular site or refinery. To our knowledge the US refining industry has not applied independent checks and balances that verify the accuracy of applying these factors. Rather, it is more often than not argued that conservative assumptions are applied to calculations, which is widely promoted as a basis by which over-reporting of emissions are claimed. But this argument is qualitative because there do not seem to be any independent studies that validate the approach. Rather, the more recent studies that appear in the literature and referenced earlier all support significant underestimation of fugitive air emissions.

5.5.2 Application of computer models and simulators

In more recent years more sophisticated tools have evolved on the commercial scene to assist refineries in making emissions determinations. These tools are software applications that rely on rigorous engineering principles and calculation methods (e.g. mass, momentum and energy transfer, thermodynamics, and chemical kinetics) to estimate emissions based on specific physical, operating, and activity parameters. A few examples are:

- GRI-GLYCalc for estimating still column off-gas emissions from glycol dehydration units (see Thompson et al., 1994);

- E&P TANK for calculating flashing and evaporation losses from production storage tanks (see DB Robinson Research Ltd., 1997);
- Tanks 4.0 for predicting evaporation losses from tanks containing stable products (i.e. based on the API correlations for evaporation losses; see US EPA, 1999);
- estimation methods of flaring and venting operations (see Canadian Association of Petroleum Producers, 2002).

There are also a variety of commercial process and emission-source simulation packages that are promoted through the Web, each making varying claims and offering different sets of calculations. A complication with all of these tools is that they generally tend to have some degree of empiricism built in. This means that they do require calibration in order to quantify the precision of calculated results. This again adds a level of cost with both labor and field measurements that need to be supported by monitoring. We do know that Tanks 3.1 and 4.0 are widely used, but even here as noted by the Shine memorandum the application of this software is not reliable largely because specific information characterizing tank construction, age, and condition are not always accurately accounted for.

We have not been able to assess how widely used these tools are, but given the fact that the industry on the whole focuses almost entirely on absolving its responsibilities by meeting the minimum requirements of statutory reporting, we doubt these more accurate tools are employed on a wide-scale basis. Facilities simply have a fallback position – that they rely on an EPA-approved calculation methodology (AP-42) and with this as a defense they claim they are using best available practices.

5.5.3 *Direct measurement techniques*

These techniques include duct or stack flow measurements, bagging, high-flow sampler, isolation flux chambers, and portable wind tunnels. The latter two methods are applicable for measuring volatilization rates from sources such as exposed oil sands, contaminated soils, and landfarm operations. In general, direct methods tend to offer the greatest potential accuracy but are only amenable to relatively simple point sources or applications where a high degree of specificity is required.

Since much of our focus has been on fugitive emissions, the subject of stack testing is not included. Direct methods are most often applied for screening purposes, but they are also critical to the support of leak detection and repair (LDAR) programs.

Source screening is performed with a portable organic compound analyzer (screening device). The portable analyzer probe opening should be placed at the leak interface of the equipment component to obtain a "screening" value. The screening value is an indication of the concentration level of any leaking material at the leak interface.

The accuracy of measurements can depend on the distance between the probe and the leak interface, as well as other parameters. The use of the leak rate correlations requires screening values gathered as closely as practicable to the leak interface.

A screening program aims to measure organic compound concentration at any potential leak point associated with a process unit. Examples of equipment types that are potential sources of leak emissions include pump seals, compressor seals, valves, pressure-relief devices, flanges, connectors, open-ended lines, agitator seals, instruments, loading arms, stuffing boxes, vents, dump lever arms, diaphragms, drains, hatches, meters, polished rods, and vents. The first step is to define the process unit boundaries and obtain a component count of the equipment that could release fugitive emissions. A process unit can be defined as the smallest set of process equipment that can operate independently and includes all operations necessary to achieve its process objective. The use of a flow diagram of the process helps to note the process streams. The actual screening data collection can be done efficiently by systematically following each stream.

Various portable organic compound detection devices are used to measure concentration levels at the equipment leak interface. The VOC detector should respond to those organic compounds being processed (determined by the response factor, RF). Both the linear response range and the measurable range of the instrument for the VOC to be measured and the calibration gas must encompass the leak definition concentration anticipated. The scale of the analyzer meter must be readable to $\pm 2.5\%$ of the specified leak definition concentration and the analyzer must be equipped with an electrically driven pump so that a continuous sample is provided at a nominal flow rate of between 0.1 and 3.0 liters per minute. The analyzer must be intrinsically safe for operation in explosive atmospheres, and it must be equipped with a probe or probe extension for sampling not to exceed 0.25 inch in outside diameter, with a single end opening for admission of sample. Examples of commonly used portable organic compound detection instruments, along with general information on their functions, are:

- Bacharach Instrument Co., Santa Clara, California Model L – detects combustible gases based on the technique of catalytic combustion over a range of 0–100% lower explosion limit (LEL). The company also markets the TLV Sniffer, which also detects combustible gases based on catalytic combustion technique over the ranges of 0–1000 and 0–10,000 ppm.
- Foxboro, S. Norwalk, Connecticut manufactures several instruments: (1) OVA-128, which detects most organic compounds based on the technique of flame ionization detection/gas chromatography (FID/GC) over the range of 0–1000 ppm. (2) OVA-108, which detects most organic compounds based on FID/GC over the range of 0–10,000 ppm. (3) Miran IBX, which detects compounds that absorb infrared radiation using the technique of non-dispersive infrared analysis (NDIR). The range of the instrument is compound specific. (4) TVA-1000, which detects most organic and inorganic compounds based on the techniques of photoionization and FID/GC. The photoionization instrument has a detection range of 0.5–2000 ppm and the FID/GC model has a range of 1–50,000 ppm.

- Health Consultants sell an instrument named Detecto- PAK III, which can detect most organic compounds using the FID/GC technique over the range of 0–10,000 ppm.
- HNU Systems, Inc. of Newton Upper Falls, Massachusetts sells an instrument called HW-101 model, which detects chlorinated hydrocarbons, aromatics, aldehydes, ketones, any substance that ultraviolet light ionizes. It operates on the technique of photoionization over the detection ranges of 0–20, 0–200, and 0–2000 ppm.
- Mine Safety Appliances Co. of Pittsburgh, Pennsylvania makes the Model 40, which detects combustible gases based on the technique of catalytic combustion over the ranges of 0–10% and 0–100% LEL.
- Survey and Analysis, Inc. of Northboro, Massachusetts makes the On Mark Model 5, which detects combustible gases based on the technique of thermal conductivity over the detection ranges of 0–5% and 0–100% LEL.
- Rae Systems of Sunnyvale, California markets the MiniRAE PGM-75K, which detects chlorinated hydrocarbons, aromatics, aldehydes, ketones, any substance that ultraviolet light ionizes. The instrument detects these compounds based on photo-ionization over the range of 0–1999 ppm.

Examples of some of these instruments are shown in Figures 5.9–5.11.

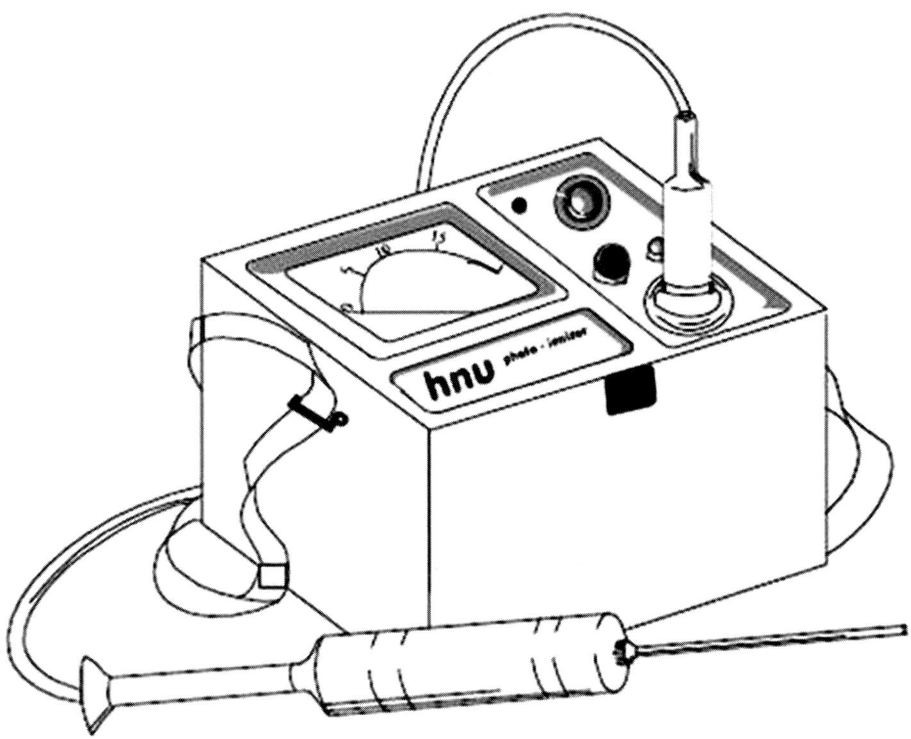

Figure 5.9 A portable HNU meter for organic compound detection.

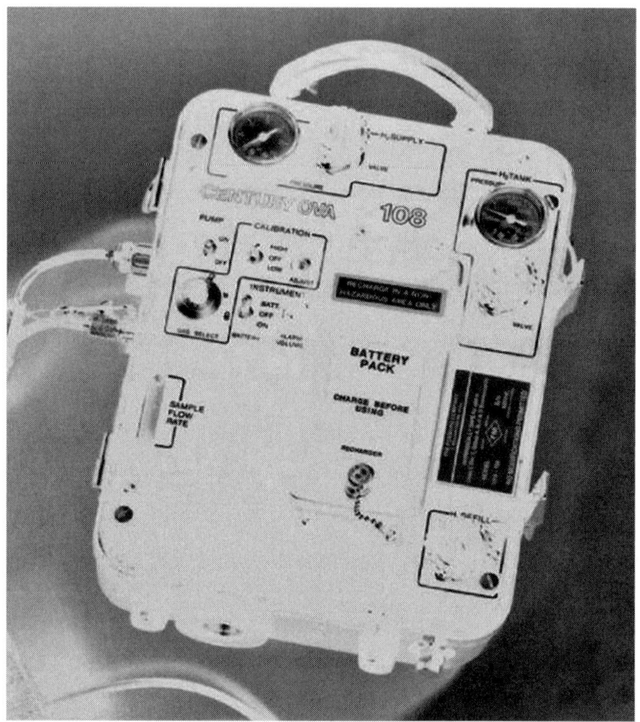

Figure 5.10 A typical OVA.

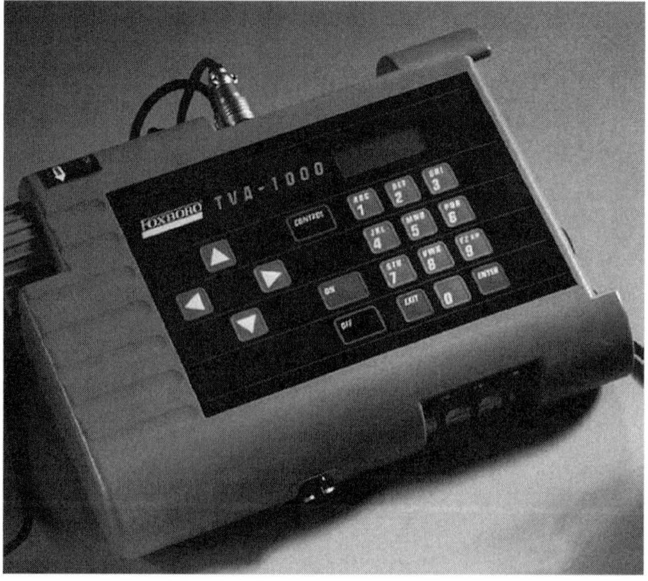

Figure 5.11 A portable TVA instrument for organic and inorganic compounds.

Most portable instruments can be integrated with data loggers to aid in the collection of screening data and in downloading the data to a laptop or desk computer. Database management programs are also available to aid in screening data inventory management and compiling emissions.

It is important to note that portable monitoring instruments do not respond to different organic compounds equally. To correct screening values to compensate for variations in a monitor's response to different compounds, response factors (RFs) have been developed. An RF relates measured concentrations to actual concentrations for specific compounds using specific instruments. Instrument suppliers generally provide this information.

Remember that these instruments only provide a snapshot of fugitive emissions. One still has to apply the field data to developing estimations of the mass emissions. Hence, we are still faced with relying on a calculation method. The accepted approach for estimating fugitive emissions from equipment leaks is to use the EPA correlation equations that relate screening values to mass emission rates. There are several methods that can be applied for emission estimation purposes with the selection based on the degree of accuracy needed. There is of course considerable cost associated with more precise methods.

Equipment leak emissions may occur randomly, intermittently, and vary in intensity over time. They are at best a "snapshot" of emissions from a given leak indicated by screening results, which are used to either develop or apply all of the approaches, and may or may not be representative of the individual leak. Since equipment is not continuously monitored, the discovery of a leak does not provide any information on how long the event has occurred. For this reason reporting requirements impose the assumption that the leak has been occurring for the entire time between test intervals. By obtaining measurements from several pieces of a given equipment type, the snapshots of individual deviations from the actual leaks are generally believed to offset one another such that the ensemble of leaks should be representative. These approaches are imperfect tools for estimating fugitive emissions from equipment leaks. We do not believe these to be the best available technologies today. Both differential absorption light detection and ranging (DIAL) technology and enhanced infrared video imaging appear to offer more precise methods for identifying emissions sources. These are described in the next section.

The effectiveness of screening tools in controlling fugitive emissions depends heavily on the frequency of sampling, and how well data obtained from field sampling is managed. As an example a fugitive emission survey study documented by the Shedgum Gas Plant in Saudi Aramco involved monitoring of up to 2000 components of flanges, valve packing, pump seals and others for fugitive emissions leaks and documenting the details in customized software (Al-Muaibid and Al-Ayadhi, 2004). The survey included the implementation of process research, flagging, tagging, documentation, monitoring, repair attempts, and report generation. About 7.5% of the surveyed component was found to be badly leaking.

In the Aramco program only selected sample components were chosen limiting the assessment to a certain area in the gas plant. The areas that were included were the Khuff Gas Processing Facilities, Hydrocarbon Condensate Stripping System, Gas Treating Units, and NGL Recovery. The implementation methodologies were as follows:

- **Process research** – this covered review of stream compositions from PFDs and P&IDs to determine applicability and the service (liquid or vapor) of the steam.
- **Flagging** – this covered identification and verification of components in the field.
- **Tagging** – this included fixing of bar-coded and numbered tags to each component that was identified during the flagging step.
- **Documenting** – this covered populating the database for all selected components by using handheld minicomputers to enter data about location, BCT number, drawing number, HAP or VOC type, and product code.
- **Monitoring** – this included monitoring all selected components by instrument detector in ppm concentration of leaking gas, utilizing a TVA-1000 (Thermo Environmental Instrument).
- **Repair attempts** – these were made in order to avoid minor leaking during the survey and then remonitoring to obtain accurate results.
- **Report generation** – reporting included generating a report of all leaking components in LEADERS Software, which is a leak detection and repair software for implementation of a fugitive emissions monitoring and modeling program.

The study utilized the following equipment and software to perform the daily routine monitoring and documentation:

- Thermo Electron Corporation TVA 1000B portable Toxic Vapor Analyzer flame ionization detector (FID);
- DAP Technology PC 9800 LS handheld computer configured with bar-code scanner;
- the LEADERS LDRS Software System;
- the EPA Method 21 Standards.

Based on the monitoring results the estimation of emission quantity has been calculated using the EPA emissions factors and based on the Integrated Method for Petroleum Process Units with 10,000 ppm PEGGED emission rate. The study showed that 153 components (7.6%) were leaking out of the 2016 components monitored. The extent of VOC emission from these leaking components was higher than had been expected.

The Aramco study is a reasonable case study because it resulted in a set of corrective actions aimed at reducing fugitive emissions on an ongoing basis. The following are the set of recommendations that the authors of the assessment implemented:

- Monitor all accessible components every 6 months and the inaccessible ones every year.
- Repair all leaking components.
- Remonitor to check effectiveness of the monitoring.
- Conduct similar surveys on the other Saudi Aramco facilities.

By way of general guidelines for best management practices, we offer the following definitions and guidelines to assist in developing screening strategies.

Definitions

These definitions are from the San Luis Obispo County Air Pollution Control standards (Rule 417 – Control of Fugitive Emissions of Volatile Organic Compounds, San Luis Obispo County Air Pollution Control District, http://www.arb.ca.gov/drdb/slo/cur.htm).

- **Background** – for gas plants, a reading expressed as methane on a portable hydrocarbon detection instrument that is taken at least 3 meters upwind from any components to be inspected and that is not influenced by any specific emission point.
- **Closed-vent system** – a system that is not open to the atmosphere and is composed of piping, connections and, if necessary, flow-inducing devices that transport gas or vapor from a piece or pieces of equipment to a vapor recovery or disposal system.
- **Component** – any valve, fitting, pump, compressor, pressure-relief device, hatch, sight-glass, meter, or open-ended lines. They are further classified as: (a) major component is any 4-inch or larger valve, any 5-hp or larger pump, any compressor, and any 4-inch or larger pressure relief device; (b) minor component is any component that is not a major component; (c) critical component is any component that would require the shutdown of the process unit if these components were shut down.
- **Fitting** – a component used to attach or connect pipes or piping details, including but not limited to flanges and threaded connections.
- **Fugitive emissions** – any hydrocarbon emissions that are released into the atmosphere from any point other than a stack, chimney, vent, or other functionally equivalent opening.
- **Hatch** – any covered opening system that provides access to a tank or container.
- **Inaccessible component** – any component located over 15 feet above ground when access is required from the ground, or any component located over 6 feet away from a platform when access is required from the platform.
- **Leak minimization** – tightening or adjusting a component for the purpose of stopping or reducing leakage to the atmosphere.
- **Major gas leak** – for gas plants, the detection of total gaseous hydrocarbons for any component in excess of 10,000 ppmv as methane above background measured.
- **Major liquid leak** – a visible mist or continuous flow of liquid.
- **Minor gas leak** – for gas plants, the detection of total gaseous hydrocarbons for any component in excess of 1,000 ppmv but not more than 10,000 ppmv as methane above background measured.
- **Minor liquid leak** – any liquid leak that is not a major leak and drips liquid organic compounds at the rate of more than three drops per minute or 1 cubic centimeter per minute.
- **Offshore oil production platform** – a unit used in the production of oil and gas that is located offshore within 3 miles of the shoreline.

- **Oil and gas production field** – a facility at which crude petroleum and natural gas production and handling are conducted, as defined by Standard Industrial Classification code number 1311, Crude Petroleum and Natural Gas.
- **Pipeline transfer station** – a facility that handles the transfer or storage of petroleum products or crude petroleum in pipelines.
- **Platform** – any raised, permanent, horizontal surface that provides access to components.
- **Polished rod stuffing box (PRSB)** – a packing device used on oil and gas production wellheads compressed around a reciprocating rod for the dual purpose of lubricating the polished rod and preventing fluid leaks.
- **Pressure-relief device (PRD)** – a pressure-relief valve or rupture disc.
- **Pressure-relief event** – a release from a pressure-release device resulting when the upstream static pressure reaches the set point of the pressure-release device. A pressure-relief event is not a leak.
- **Pressure-relief valve (PRV)** – a valve that is automatically actuated by upstream static pressure and used for safety or emergency purposes.
- **Process unit** – a manufacturing process that is independent of other processes and is continuous when supplied with a constant feed of raw material and sufficient storage facilities for the final product.
- **Process unit shutdown** – a work practice or operational procedure that stops production from a process unit or part of a process unit. An unscheduled work practice or operational procedure that stops production from a process unit or part of a process unit for less than 24 hours is not a process unit shutdown. The use of spare equipment and technically feasible bypassing of equipment without stopping production are not process unit shutdowns.
- **Pump** – a device used to provide energy for transferring a liquid or gas/liquid mixture through a piping system from a source to a receiver.
- **Refinery** – a facility that processes petroleum, as defined by the Standard Industrial Classification Code number 2911, Petroleum Refining.
- **Repair** – Any corrective action for the purpose of eliminating leaks.
- **Rupture disc** – a diaphragm held between flanges for the purpose of isolating a volatile organic compound from the atmosphere or from a downstream pressure-relief valve.
- **Seal** – packing gland or other material compressed around a moving rod, shaft, or stem to prevent the escape of gas or liquid.
- **Unmanned facility** – a remote facility that has no permanent sited personnel and is greater than 5 miles from the closest manned facility, operated by the same company or corporation.
- **Unsafe to monitor components** – components installed at locations that would prevent the safe inspection or repair of components as defined by OSHA standards, the provisions for worker safety found in 29CFR1910, or written owner-supplied criteria.
- **Valve** – a device that regulates or isolates the fluid flow in a pipe, tube, or conduit by means of an external actuator.
- **Vapor control system** – any system not open to the atmosphere intended to collect and reduce volatile organic compound emissions to the atmosphere and composed of piping, connections and, if necessary, flow-inducing devices that transport gas or vapor from a piece or pieces of equipment to a vapor recovery or disposal system.

- **Volatile organic compound (VOC)** – any compound containing at least one atom of carbon, except exempt compounds.

Best practices

The following best practices were summarized from the San Luis Obispo County Air Pollution Control District guidelines.

Inspection frequencies

- A leak is any liquid leak, a visual or audible vapor leak, the presence of bubbles using soap solutions, or a leak identified by the use of a vapor analyzer.
- Any vapor leak that is identified during the inspection of components should be measured to quantify emission concentrations according to EPA Reference Method 21.
- All pumps, compressors, and PRVs should be inspected for leaks once during every 8-hour period or, with written approval once during every operating shift, except for components located at manned and unmanned oil and gas production fields and pipeline transfer stations.
- All pumps, compressors, PRVs, and PRSBs located at manned oil and gas production fields and pipeline transfer stations should be inspected for leaks once per day and components located at unmanned facilities should be inspected once per week.
- All components should be inspected quarterly according to EPA Method 21.
- All inaccessible components should be inspected annually according to EPA Method 21.
- All fittings should be inspected for leaks according to EPA Method 21 immediately after being placed in service and semi-annually thereafter.
- All critical and unsafe to monitor components should be inspected in accordance with an approved inspection plan.
- A pressure-relief valve should be inspected according to EPA Reference Method 21 within 3 calendar days after every pressure relief.
- The inspection frequency for components, except pumps, compressors, PRVs, and PRSBs, may change to an annual inspection, provided all of the following conditions are met:
  - All components at the facility have been successfully operated and maintained with no liquid leaks and no major gas leaks exceeding 0.5% of the total components inspected per inspection period for 12 consecutive months. Leaks from PRSBs should not be included in the total count of leaking components.
  - The above should be substantiated by documentation and written approval.
- Any approved annual inspection frequency should revert to the inspection frequencies specified earlier should any liquid leaks and major gas leaks exceed 0.5% of the total components inspected per inspection period.
- All leaking components should be affixed with brightly colored, weatherproof tags showing the date of leak detection. The tags should remain in place until the components are repaired and reinspected.

Equipment repairs

- The requirements should apply in all situations when a leak is detected.

- All noncritical components should be successfully repaired or replaced within the following time periods after detection of the leak according to [Table 5.9].
- Leaks from components should be immediately minimized to the extent possible to stop or reduce leakage to the atmosphere.
- All leaks from critical and unsafe to monitor components should be minimized to the extent possible and should be replaced with best available control technology (BACT) equipment within 1 year or during the next process unit shutdown, not to exceed 2 years.
- Any repaired or replaced component should be reinspected in accordance with EPA Method 21 by the operator within 30 days of the repair or replacement.
- A component or part that incurs five repair actions for a major gas or liquid leak within a continuous 12-month period should be replaced with BACT equipment.
- Open-ended lines and valves located at the end of lines should be sealed with a blind flange, plug, cap, or a second closed valve at all times except during operation. Operation includes draining or degassing operations, connection of temporary process equipment, sampling of process streams, emergency venting, and other normal operational needs.
- Hatches should be closed at all times except during sampling, adding process material, or attended maintenance operations.

Equipment identification

- All major and critical components should be physically identified (clearly and visibly) for inspection, repair, replacement, and recordkeeping purposes.
- All major, critical, inaccessible, and unsafe to monitor components except fittings should be clearly identified in diagrams for inspection, repair, replacement, and recordkeeping purposes.

Test methods

- Measurements of total gaseous hydrocarbon leak concentrations should be conducted according to EPA Reference Method 21.
- The volatile organic compound content of fluids should be determined using ASTM Methods E 168-88, E 169-87, or E 260-85.

Table 5.9 Suggested time periods between repairs

Type of leak	Time period (days)[a]	
	Onshore	Offshore
Minor gas leak	14	14
Major gas leak	5	5
Major gas leak over 50,000 ppm	1[b,c]	5
Major liquid leak	1[b,c]	5
Minor liquid leak	2[b,c]	5

[a]Day means a 24-hour period from the time of leak detection.
[b]Unless prohibited by California OSHA standards or 29 CFR1910.
[c]Components located at oil and gas production facilities or pipeline transfer stations should be repaired within 2 days.

- Determination of exempt compounds should be performed in accordance with ASTM D4457-85. For exempt compounds where no reference test method is available, a facility requesting the exemption should provide appropriate test methods approved by the US Environmental Protection Agency.
- Determination of evaporated compounds of liquids should be performed in accordance with ASTM D 86-82.
- Determination of the API gravity of crude oil should be performed in accordance with ASTM Method D 287.

Recordkeeping

- Each facility operator should maintain an up-to-date inspection log containing, at a minimum, the following:
  - Name, location, type of components, and description of any unit where leaking components are found.
  - Date of leak detection, emission level (ppmv) of leak, and method of leak detection.
  - Date and emission level of recheck after leak is repaired.
  - Total number of components inspected, and total number and percentage of leaking components found by component types.
  - Records of leaks detected by a quarterly or annual operator inspection, and each subsequent repair and reinspection, should be submitted to the appropriate regulatory agency.
  - All records of operator inspection and repair should be maintained at the facility for the previous 2-year period and made available to the regulatory agency upon request.
  - The facility should maintain a report on the previous year's inspection and maintenance activities, which (i) summarizes the inspection log entries, and (ii) lists all leaking components identified that were not repaired within 15 days and all leaking components awaiting a unit turnaround for repairs.

5.5.4 *Indirect measurement techniques*

A technique known as differential absorption light detection and ranging (DIAL) has been used to remotely measure concentration profiles of hydrocarbons for refinery surveys in Europe for over 15 years. The technique is non-invasive and single-ended, and gives concentration profiles and mass emissions of various species in the area being surveyed.

The measurement relies on a "fingerprint" absorption spectrum of each molecule. An absorption measurement is made with laser light, at a peak of absorption (lambda-on) and at a trough (lambda-off), resulting in a differential signal. The differential nature of the signal is the underlying principle of the measurement process. Figures 5.12 and 5.13 provide simplified diagrams of the method. As shown in Figure 5.12 a pulsed laser beam is sent out into the test area and small proportions of the light are backscattered by particles along the beam path to a sensitive detector that is illustrated in Figure 5.13. Dust particles and aerosols are used as weak reflectors. The laser light is in short pulses and time resolution of the backscattered light (along with the

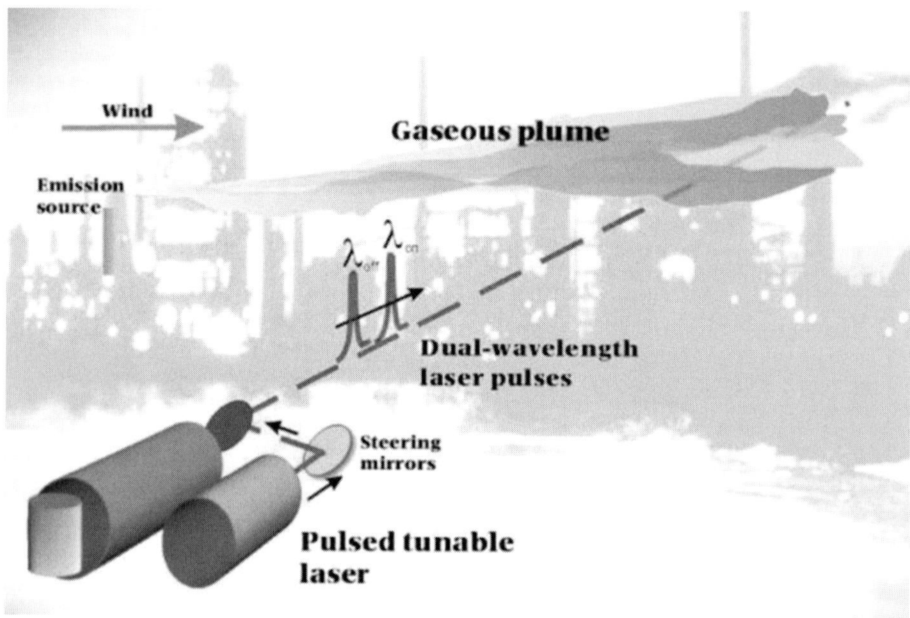

Figure 5.12 DIAL method output signal. From Spectrasyne web page, http://www. spectrasyne.ltd.uk/html/technique.html

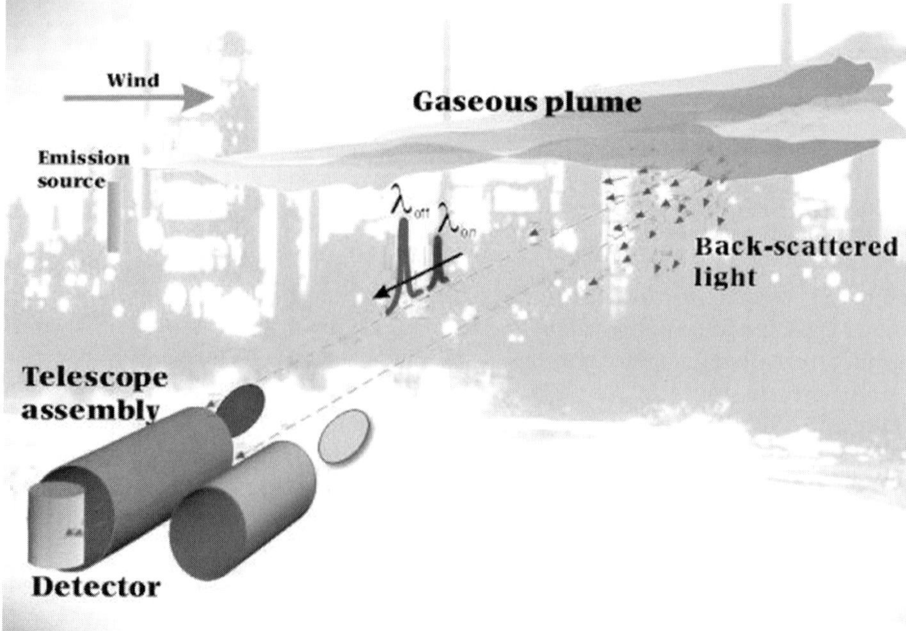

Figure 5.13 Return signal. From Spectrasyne web page, http://www.spectrasyne.ltd.uk/ html/technique.html

speed of light) gives range resolution as in a LIDAR (light detection and ranging).

For concentration measurements the system relies on a differential return from two closely spaced wavelengths, only one of which is absorbed strongly by the target gas. The size of the differential return signal at different distances along the laser beam path provides a sensing of the contaminant's concentration. The concentrations are converted into mass emissions by making a series of scans with the DIAL along different lines within a plume and combining these with meteorological data in a software package. These measurements are then used to produce a mass emission profile for the entire refinery.

The equipment used to measure the emissions is comprised of lasers, computers, and associated apparatus housed in a mobile unit known as the Environmental Surveying System (ESS). The ESS contains two complete Nd:YAG pumped, dual-wavelength dye lasers to provide the multi-wavelength source for DIAL measurements. The laser systems incorporate frequency doubling and mixing accessories to give a range of ultraviolet (UV) and infrared (IR) wavelengths that augment the visible and near-IR spectrum produced by the dye lasers alone.

The output beams from the lasers are directed into the measurement area by means of computer-controlled steering mirrors and collection of the returning signal is via a Cassegrain telescope. Data processing is performed via a high-speed data communication network, which has been developed in parallel with a unique software package.

The ESS is equipped with an extendible meteorological mast and a number of mobile, telemetric stations that are used to measure wind speed and direction, temperature and humidity. These measurements are used in conjunction with the DIAL concentration measurements to compute mass emissions. While there are other methods described in the literature, DIAL is the only measurement technique available that enables mass emission fluxes to be obtained directly.

Meteorological ("met") measurements play an important role in the quantification of the analysis. A 14.5-m mast mounted in the DIAL vehicle measures the free-air wind speed and direction. In addition, remote anemometers and wind vanes are deployed along the DIAL scan plane at various heights and the results are sent back at regular intervals via radio links to the DIAL vehicle, which uses the data in "real-time" to calculate fluxes.

The system has a maximum working range of over 1 km, a range resolution of a few meters, and a detection sensitivity of a few parts in 10^8. Current target species include CH_4 and hydrocarbons in the IR, benzene, toluene, NO and NO_2, benzene and toluene in the UV and visible, and other compounds.

The typical refinery mixtures of fugitive species are measured in a spectral region around 3 μm wavelength where absorption of most of the light hydrocarbons overlap. This region thus provides the opportunity for the DIAL system to identify the majority of atmospheric species in a single measurement.

Sorption tubes are also deployed during DIAL measurements to collect a range of aromatic species. When analyzed and correlated with the DIAL-measured aromatic species, a mass emission can be calculated for a range of other species (e.g. xylenes, ethylbenzene).

The technology was perfected and is marketed by a company called Spectrasyne, which was formed by a management buyout from British Petroleum. Whilst working for BP Research, two of the current Spectrasyne directors undertook a DTI/Industry collaborative research and development program that enabled the development of DIAL. The company has deployed its ESS mobile unit on numerous refinery surveys throughout Europe since the mid-1990s.

In a first-of-a-kind demonstration in a North American refinery, Environment Canada (Chambers and Strosher, 2006) completed a project in 2005 in which the DIAL technology was demonstrated along with infrared video camera imaging, which is a technique that appears to be effective for screening for fugitive emissions from leaks. The gas leak imaging cameras with modified infrared video cameras visually detect hydrocarbon plumes from leaking equipment. This latter technique seems well suited for real-time monitoring as well as more accurate and less time- consuming screening for fugitive emissions.

The Alberta refinery DIAL trial was conducted over a 10-day period. The technique was used to quantify emissions of methane, C_{2+} hydrocarbons (i.e. alkane hydrocarbons ethane and larger), and benzene. The monitoring apportioned the hydrocarbon emissions to specific target areas within the refinery. The gas leak imaging technique was demonstrated over a 5-day trial period that coincided with the DIAL measurements. A review of the published work supports that both techniques complement each other, with the imaging technology most beneficial for identifying leaks that can be immediately repaired and for directing DIAL measurements to problem areas.

The total fugitive emissions were found to be 1240 kg/h of C_{2+} hydrocarbons, 200 kg/h of methane, and 5 kg/h of benzene. The reader should bear in mind that these are not total facility emissions but rather they are emissions from select areas of the refinery. Aside from a community exposure standpoint, the investigators attempted to correlate the mass emissions with monetary losses. It was reported that these emissions were equivalent to $3.2 million per year in material losses based on a crude value of $40 per barrel. Hence, the DIAL technology lends itself to being an important pollution prevention tool that has the potential to identify and help capture direct savings for refinery operations.

The Alberta DIAL trial resulted in some additional important findings. Firstly, it showed that emissions from storage tanks accounted for more than 50% of the total fugitive emissions measured of both C_{2+} hydrocarbons and benzene. Secondly, when the mass measurements were compared to those calculated from EPA emission factors, the DIAL measurements not only showed an entirely different distribution of source emissions, but overall much greater levels of mass emissions.

5.5.5 Activity data

While we are critical of the use of emission factors, we recognize that these types of calculations are still useful from both a screening standpoint and in lieu of other more precise methods that regulations do not mandate be adopted for monitoring and reporting purposes at this point in time. In order to achieve some degree of confidence in calculated emissions it is critical that accurate activity data be assembled. Activity data may include production statistics, infrastructure data (e.g. inventories of facilities/installations, process units, pipelines, and equipment components), and reported emissions from spills, accidental releases, and third-party damages. The basic activity data required regardless of the assessment method are extensive for a refinery or a gas processing plant. Table 5.10 provides a summary of the type of data that must be assembled in order to prepare an accurate emissions inventory.

Table 5.10 List of activity data required for an emissions inventory

Primary source category	Minimum required activity data
All	Oil and gas throughputs
Oil systems	Gas-to-oil ratios; flared and vented volumes; conserved gas volumes; reinjected gas volumes; utilized gas volumes; gas compositions
Process venting/flaring	Reported volumes; gas compositions; proration factors for splitting venting from flaring
Storage losses	Solution gas factors; liquid throughputs; tank sizes; vapor compositions
Equipment leaks	Facility/installation counts by type; processes used at each facility; equipment component schedules by type of process unit; gas/vapor compositions
Gas-operated devices	Schedule of gas-operated devices by type of process unit; gas consumption factors; type of supply medium; gas composition
Accidental releases and third-party damages	Incident reports/summaries
Gas migration to the surface and surface casing vent blows	Average emission factors and numbers of wells
Drilling	Number of wells drilled; reported vented/flared volumes from drill stem tests; typical emissions from mud tanks
Well servicing	Tally of servicing events by types
Pipeline leaks	Type of piping material; length of pipeline
Exposed oil sands/oil shale	Exposed surface area; average emission factors

In applying more refined analyses to developing emissions inventories the activity data required may include the following:

- process operating conditions (e.g. gas compositions, temperatures, pressures and flows);
- maintenance records;
- accident reports;
- tallies and details of blowdown events, compressor starts, and purging activities;
- inventories of gas-operated devices that use natural gas as the supply medium (e.g. instrument control loops, chemical injection pumps, automatic samplers);
- facility or installation counts by type;
- process unit counts by type;
- gas-to-oil ratios;
- number and types of wells drilled, tested, and serviced;
- number of pipeline tie-ins resulting in blowdown and purging events;
- pigging frequencies per pipeline system;
- production rates;
- vented volumes;
- flared volumes;
- population, sizes, and service of storage tanks;
- equipment component counts on a process unit and facility basis;
- length of each pipeline and type of pipe material used (e.g. steel, cast iron, aluminum, or plastic);
- emissions control measures;
- sweet, sour, or odorized service;
- operating practices (depressurization of idle compressors, flaring rather than venting, etc.).

Some important considerations in obtaining and using refined activity data include:

- The required data for a refined assessment are often difficult and costly to obtain or are simply unavailable. Consequently, some assumptions may be needed to bridge certain information gaps.
- For large oil and gas complexes, the emission inventories will be susceptible to significant errors due to missed or unaccounted sources. To minimize such errors it is important to obtain active industry involvement in the preparation and refinement of these emission inventories. Difficulties that may be encountered in attempting to utilize available data are as follows:
  - converting electronic data to a consistent or convenient format;
  - reliable and accurate data entry (particularly where large amounts of information are involved);
  - verification of database accuracies and completeness;
  - establishing the existence or availability of information (e.g. useful statistics may be maintained for reasons such as taxation, equipment maintenance, design documentation, property insurance policies, financial accounting, etc., and not be known to those charged with developing and maintaining the emissions inventory).

- The radiative forcing of vented waste gas volumes may be 7–12 times greater than that for equivalent flared volumes (depending on the gas composition and flaring efficiency). Consequently, these two activities must be considered separately. However, statistics on venting and flaring activities usually are available only as a combined volume and often are simply reported as flared gas. Many production accounting systems do not track flared volumes and vented volumes separately, or the ability to do so is not utilized. Consequently, all reported flared and vented volumes should be scrutinized carefully and some values may need to be adjusted.
- Except for acid gas flares and some continuous waste gas flares, flare and vent systems normally are not equipped with flow recorders. Consequently, reliable estimates of flared volumes can be difficult to obtain.
- Problems such as simmering or leaking pressure-relief valves can be significant contributors to total flared or vented volumes and go unreported.
- Depending on the accounting procedures of individual companies, reported venting volumes may not include solution gas emissions from production storage tanks, vented volumes by gas-operated devices and compressor starts, blowdown volumes from maintenance and repair activities, and still-column off-gas emissions from glycol dehydrators.
- The concentrations of CH_4, H_2S, and CO_2 naturally present in the produced hydrocarbons may vary dramatically from one field to the next.

5.5.6 Production statistics

The following are some general guidelines on the production statistics that are needed to develop a basis for emissions estimates:

- The production statistics should be disaggregated to capture changes in throughputs (due to imports, exports, reprocessing, withdrawals, etc.) in passing through oil and gas systems.
- If data are gathered from several different sources, it must be ensured there is no double or missed counting of emissions due to differences in terminology and classifications.
- Uncertainty will exist if there is any inherent bias in the original measurement results (for example, sales meters are often designed to err in favor of the customer, and liquid handling systems will have a negative bias due to evaporation losses). Random metering and accounting errors may be significant and should be verified.
- Reported vented and flare volumes are highly suspect since these values are sometimes estimates and not based on actual measurement results. Additionally, the values are often aggregated and simply reported as flared volumes at some facilities.
- Operating practices of each segment of a refinery should be reviewed to determine if the reported volumes are actually vented or flared, or to develop appropriate proration factors. Furthermore, audits or reviews of each portion of a refinery or gas plant should be conducted to determine if all vented/flared volumes are actually reported (for example, solution gas emissions from storage tanks and treaters, emergency flaring/venting, leakage into vent/flare systems, and blowdown and purging volumes may not necessarily be accounted for).

- Some production statistics may be reported in units of energy (based on their heating value) and need to be converted to a volume basis, or vice versa, for application of the available emission factors. Typically, where production values are expressed in units of energy, it is in terms of the gross (or higher) heating value of the product. However, where emission factors are expressed on an energy basis it is sometimes in terms of the net (or lower) heating value of the product (especially factors provided by equipment manufacturers). Accordingly, it must be ensured that the emission factors and activity data are on a consistent basis.

5.5.7   Data infrastructure

While one would expect that infrastructure data are readily available on a facility-by-facility basis, especially since refineries and gas production plants in the USA are mature, the example of PXP's poor component inventory practices points towards a concern. For those companies that are diligent in maintaining accurate component and equipment inventories, computerized inspection-and-maintenance information management systems are definitely the best management tools relied upon. These systems are a reliable means of counting major equipment units (compressor units, process heaters and boilers, etc.) at facilities. Also, we have observed that some departments within overseas refineries maintain databases of certain types of equipment or facilities for their own specific needs (tax accounting, production accounting, insurance records, quality control programs, safety auditing, license renewals, etc.). Some efforts should be made to identify these potentially useful pools of information. Most certainly facilities that have achieved ISO 14001 registration follow such practices.

Component counts by type of process unit will certainly vary dramatically between refineries due to differences in design and operating practices. Relying on values reported in the general literature is at best only useful for comparative purposes. As a matter of proper due diligence facilities should develop their own inventories and maintain them. We recommend that this be done by an external firm to ensure accuracy and to avoid conflicts of interest when such information is relied upon for compliance reporting purposes such as preparing compliance emissions reporting.

The use of consistent terminology and clear definitions is critical in developing counts of facilities and equipment components, and to allow any meaningful comparisons of the results and in applying the data to developing emissions estimates.

5.5.8   Additional comments on accuracy and confidence limits

In Section 5.3 we discussed inherent flaws that result in biased reporting of fugitive emissions. These criticisms largely fall on the general assumptions that are applied by the industry sector to the use of emission factors. But let us assume that our criticisms and those of the EPA are not valid. There are other potential sources of uncertainties in creating inventories of fugitive emissions that contribute to biased reporting. These include:

- measurement errors;
- extrapolation errors;
- inherent uncertainties of the selected estimation techniques;
- missing or incomplete information regarding the source population and activity levels;
- poor understanding of temporal and seasonal variations in the sources;
- over- or under-accounting due to confusion or inconsistencies in category divisions and source definitions;
- misapplication of activity data or emission factors;
- errors in reported activity data;
- missed accounting of intermediate transfer operations and reprocessing activities (e.g. repeat dehydration of gas streams [in the field, at the plant, and following storage], treating of slop and foreign oil receipts) due to poor or no documentation of such activities;
- variances in the effectiveness of control devices and missed accounting of control measures;
- data entry and calculation errors.

Due to the complexity of the oil and gas industry it is difficult to quantify the net uncertainties in the overall inventories, emission factors, and activity data, but certainly not impossible. While some semiquantitative analyses have been conducted, a more thorough quantitative analysis is unquestionably warranted and should be performed if a facility certifies its emissions reporting as is required by regulatory agencies. We have never seen an example where prepared emissions inventories were accompanied by stated confidence limits. But facilities do have corporate representatives sign off on emissions reports under penalty of law that their calculated emissions are based on best available information and are true and accurate. Why shouldn't such certification be accompanied by a qualification as to how confident or accurate the mass emissions calculations are?

There is a basis for establishing confidence limits, although a standardized approach seems to be lacking in the literature. As a starting basis, Picard (2006) has noted that we know in general terms that high-quality refined emissions factors for most gases may be expected to have errors of the order of $\pm 25\%$, and that factors based on stoichiometric ratios may be much better (e.g. errors of $\pm 10\%$). We also know that gas compositions are usually accurate to within $\pm 5\%$ on individual components. Typically, flow rates have errors of $\pm 3\%$ or less for sales volumes, and $\pm 15\%$ or more for other volumes. A high-quality bottom-up inventory accounting of fugitive methane losses from either oil or gas activities might be expected to have errors of $\pm 25\%$ to as much as 50% according to Picard. In comparison, default production-based emission factors for methane losses may easily be in error by an order of magnitude or more. Inventories of fugitive CH_4 and CO_2 emissions from venting and flaring activities will be quite reliable if the raw gaseous composition and actual vented and flared volumes are accurately known. Estimates of fugitive N_2O emissions will be least reliable. Estimates of emission reductions from individual control actions may be accurate within a small percentage of $\pm 25\%$ depending on the number of subsystems or sources considered.

5.6 Closing remarks

Of the top 10 most frequently reported chemicals in the TRI list, the prevalence of volatile chemicals emphasizes the air intensive toxic chemical loading of the refining industry. Nine of the 10 most commonly reported toxic chemicals are highly volatile. Seven of the 10 are aromatic hydrocarbons that include benzene, toluene, ethylbenzene, xylene, cyclohexane, 1,2,4-trimethylbenzene, and ethylbenzene. The aromatic hydrocarbons are highly volatile and comprise a portion of both crude oil and many of the finished petroleum products.

The most recent studies support that these compounds as fugitive emissions are significantly under-reported in the TRI and that the industry relies almost entirely on calculation methods that are riddled with many sources of uncertainty. In North America the industry relies almost entirely on calculation methods that have not been updated in decades, are not verified by independent monitoring, and are absent of any statements of confidence limits.

In Section 5.3.2 we provided an example in which the results of two sets of calculations for a tank roof landing incident resulted in several orders of magnitude difference in VOC emissions. Both sets of calculations relied on EPA-approved methodology, but the selection of one method over another depends on the proper characterization of the incident. As noted by the EPA, there are many instances where episodic events resulting in unplanned releases are mischaracterized. The EPA has also made note of the fact that, often, wrong assumptions and improper input data are used in making fugitive emissions calculations. While we certainly do not believe that the industry sector on the whole intentionally understates its emissions, the fact remains that when the polluter is faced with a choice between a conservative set of assumptions that result in lower emissions reporting versus one that results in a higher estimate of emissions, the propensity to select the more conservative approach is greater. Relying entirely on calculations procedures that are prepared solely by the polluter and not verified with actual monitoring is a conflict of interest. The assertions that conservatism is built into calculation methods applied in current practices is not substantiated and most certainly discredited by recent testing using advanced monitoring technologies.

The limited monitoring that has been done in North America with techniques used in Europe for more than a decade and a half has shown that refineries are under-reporting their fugitive emissions by several orders of magnitude. This disclosure has been met by controversy and criticisms on the part of the American Petroleum Institute, who have argued that the DIAL measurements, for one, are inaccurate and not substantiated. A central point in the API's argument is that the tests conducted in Alberta were short term and that time-averaged emissions would be expected to be more consistent with published emission factors. We find this argument to be unqualified and one that misses a critical point. The DIAL measurements along with the EPA's assessment of fugitive emissions resulting from turnaround and upset and transient operational events clearly show significant excursions that are not accounted for in overall emissions reporting. Even if average yearly emissions are reasonably represented by average emission factor calculations, both workers and communities are placed at undue risks from exposures to episodic events that could be identified and more rapidly controlled through monitoring. The entire foundation of environmental regulations is to err on the side of conservatism. Current practices for preparing emissions inventories are focused on meeting statutory reporting requirements and simply sidestep the intent of environmental legislation, which is to protect human lives and the environment.

Further, monitoring makes more sense from a cost–benefit standpoint. The Alberta refinery tests conducted took the effort to examine the value of emissions. Pollution prevention focuses on reducing all forms of waste and inefficiency, not just pollution. In fact, we may argue that all forms of waste and inefficiencies are pollution since they have no value and cost any business money. Fugitive emissions represent sizeable direct product and monetary losses for refineries. But they also represent indirect costs to companies because of the increased costs to the healthcare system in the USA as well as the costs of legal actions.

References

Al-Muaibid, J.B., Al-Ayadhi, A.S., 2004. Fugitive Emission Monitoring Program Study at Shedgum Gas Plant, Saudi Aramco; http://unpan1.un.org/intradoc/groups/public/documents/ARADO/UNPAN020865.pdf.

Arora, S., Cason, T.N., 1996. Why Do Firms Volunteer to Exceed Environmental Regulations? Understanding Participation in EPA's 33/50 Program. Land Economics 72 (4), 413–432.

Baker v. Chevron USA, Inc., Case No. 1:05 CV 227, United States District Court for the Southern District of Ohio, Cincinnati Division.

Bates Stamp Document AIR000043 in the matter of Baker v. Chevron USA, Inc., Case No. 1:05 CV 227, United States District Court for the Southern District of Ohio, Cincinnati Division.

California Air Pollution Control Offices Association (CAPCOA), 1999. California Implementation Guidelines for Estimating Mass Emissions of Fugitive Hydrocarbon Leaks at Petroleum Facilities, February.

Canadian Association of Petroleum Producers, 2002. Estimation of Flaring and Venting Volumes from Upstream Oil and Gas Facilities. Publication 2002-009, May. Canadian Association of Petroleum Producers, Calgary, Alberta; available at http://www.capp.ca/raw.asp?x=1&;dt=PDF&dn=38234.

Chambers, A., Strosher, M., 2006. Refinery Demonstration of Optical Technologies for Measurement of Fugitive Emissions and for Leak Detection. Prepared for Environment of Canada – Ontario Ministry of the Environment and Alberta Research Canada Inc. Alberta, Canada, Edmonton. Project No. CEM 9643-2006, 31 March.

De Marchi, S., Hamilton, J.T., 2006. Assessing the Accuracy of Self-reported Data: An Evaluation of the Toxics Release Inventory. Journal of Risk Uncertainty 32, 57–76.

GRI Canada, 1998.

Hensley, L. v. Hoss, P.T. Plains Exploration and Production, Superior Court of California, Case No. SC094173.

Khanna, M., Damon, L.A., 1999. EPA's Voluntary 33/50 Program: Impact on Toxic Releases and Economic Performance of Firms. Journal of Environmental Economics and Management 37, 1–25.

Khanna, M., Quimio, W.R.H., Bojilova, D., 1998. Toxics Release Information: A Policy Tool for Environmental Protection. Journal of Environmental Economics and Management 36, 243–266.

Konar, S., Cohen, M.A., 1997. Information as Regulation: The Effect of Community Right to Know Laws on Toxic Emissions. Journal of Environmental Economics and Management 32, 109–124.

Picard, D., 2006. Fugitive Emissions from Oil and Natural Gas Activities, Good Practice Guidance and Uncertainty Management in National Greenhouse Gas Inventories; article on the Web – see http://www.ipcc-nggip.iges.or.jp/public/gp/bgp/2_6_Fugitive_Emissions_from_Oil_and_Natural_Gas.pdf.

Poje, G.V., Horowitz, D., 1990. Phantom Reductions: Tracking Toxic Trends. National Wildlife Federation, Washington, DC.

DB Robinson Research Ltd., 1997. A Program for Estimating Emissions from Hydrocarbon Production Tanks. Available from American Petroleum Institute (API). E&P TANK Version 1.0.

Rule 417 – Control of Fugitive Emissions of Volatile Organic Compounds, San Luis Obispo County Air Pollution Control District, http://www.arb.ca.gov/drdb/slo/cur.htm.

Shine, B., 2007. Potential Low Bias of Reported VOC Emissions from the Petroleum Refining Industry. Technical Memorandum to EPA. Docket No. EPA-HQ-OAR-2003-0146, PA/SPPD, 27 July.

Thompson, P.A., Espenscheid, A.P., Berry, C.A., Myers, D.B., 1994. Technical Reference Manual for GRI-GLYCalc: A Program for Estimating Emissions from Glycol Dehydration of Natural Gas. Version 2.0. Prepared by Radian Corporation for Gas Research Institute, Chicago, IL.

US Environmental Protection Agency (EPA), 1990. Analysis of Non-Respondentsto Section 313 of the Emergency Planning and Community Right-to-Know Act. Study prepared by Abt Associates for EPA Office of Toxic Substances.

US Environmental Protection Agency (EPA), 1995. Profile of the Petroleum Refining Industry, Enforcement and Compliance Assurance, EPA 310-R-95–013, September.

US Environmental Protection Agency (EPA), 1995. Profile of the Organic Chemical Industry, SIC 286 EPA/310-R-95-012, EPA Office of Compliance Sector Notebook Project Office of Compliance, Office of Enforcement and Compliance Assurance, Wahington.

US Environmental Protection Agency (EPA), 1999. Tanks 4.0. A program for estimating normal evaporation losses (i.e., standing and breathing losses) from production storage tanks. Available from US EPA (www.epa.gov/ttn/chief).

US Senate Committee on Environment and Public Works, 1990. p. 132.

Waxman report, Oil Refineries Fail to Report Millions of Pounds of Harmful Emissions, 1999. Report prepared for Rep. Henry A. Waxman, Minority Staff: Special Investigations Division. Committee on Government Reform, US House of Representatives. November.

Whetzel, C., 2003. South Coast Air District Seeks $319 for Violations at Los Angeles Area Refinery. The Bureau of National Affairs, Inc, Washington, DC.

6 Guidelines for cleaner production

6.1 Introduction

This chapter provides guidance on pollution prevention and cleaner production technologies. The guidelines represent state-of-the-art thinking on how to reduce pollution emissions. Not all of the methods are suitable for all refineries. They must be examined on a case-by-case basis with an eye on the cost benefits. The guidelines are intended to protect human health, reduce mass loadings to the environment, based on commercially proven technologies, generally believed to be cost-effective, follow current regulatory trends, and promote good industrial practices. These in turn offer the potential for greater productivity, increased energy efficiency, and improved environmental performance.

6.2 Best practices

The reader may refer to the US Environmental Protection Agency (EPA), *Profile of the Petroleum Refining Industry*, EPA 310-R-95-013, September 1995 (US Government Printing Office, Washington, DC) for a general discussion of baseline operations of refineries. Pollution associated with petroleum refining includes volatile organic compounds (VOCs), carbon monoxide (CO), sulfur oxides (SO_x), nitrogen oxides (NO_x), particulates, ammonia (NH_3), hydrogen sulfide (H_2S), metals, spent acids, and numerous toxic organic compounds. Sulfur and metals result from the impurities in crude oil. The other wastes represent losses of inputs and final product. These wastes are discharged in various forms such as air emissions, wastewater, or solid waste. All of these wastes require treatment and containment. Air emissions are more difficult to capture than wastewater or solid waste. Air emissions are the largest source of untreated wastes released to the environment.

The following sections describe general practices and offer some guidance on best practices aimed at improving environmental performance.

6.2.1 Air emissions

Air emissions are comprised of point and nonpoint sources. Point sources are emissions that are discharged from stacks and flares and, thus, can be monitored and treated. Nonpoint sources are fugitive emissions that are difficult to locate and capture. Fugitive emissions occur throughout refineries and arise from leaks from the thousands of valves, pumps, tanks, pressure-relief valves, flanges, etc. While individual leaks are typically small, the sum of all fugitive leaks at a refinery can be one of its largest emission sources. Fugitive emissions may also consist of area sources; for example, land areas that have been contaminated from spills and poor housekeeping are sources of airborne dusts that are contaminated by oil and refined products.

Refineries employ large numbers of process heaters to heat process streams or to generate steam (boilers) for heating or steam stripping. These are sources of SO_x, NO_x, CO, particulates, and hydrocarbon emissions. When operating properly and when burning cleaner fuels such as refinery fuel gas, fuel oil or natural gas, these emissions may be relatively low. If, however, combustion is not complete, or heaters are fueled with refinery fuel pitch or residuals, emissions can be significant.

The majority of gas streams exiting each refinery process contain varying amounts of refinery fuel gas, hydrogen sulfide and ammonia. These streams are collected and sent to the gas treatment and sulfur recovery units to recover the refinery fuel gas and sulfur. Emissions from the sulfur recovery unit (SRU) contain H_2S, SO_x, and NO_x. Other emissions sources from refinery processes arise from periodic regeneration of catalysts. These processes generate streams that may contain relatively high levels of carbon monoxide, particulates, and VOCs. Before being discharged to the atmosphere, these off-gas streams may be treated first through a carbon monoxide boiler to burn carbon monoxide and VOCs, and then sent through an electrostatic precipitator or cyclone separator to remove particulate matter (PM).

Sulfur is generally removed from a number of process off-gas streams (known as sour gas) in order to meet the Clean Air Act SO_x emissions limits and to recover saleable elemental sulfur. Process off-gas streams, which are sour gas from the coker, catalytic cracking unit, hydrotreating units, and hydroprocessing units, generally contain high concentrations of hydrogen sulfide mixed with light refinery fuel gases. Before elemental sulfur can be recovered, the fuel gases, which are primarily methane and ethane, first need to be separated from the hydrogen sulfide. This is accomplished by dissolving the hydrogen sulfide in a chemical solvent. The solvents most commonly used are amines, such as diethanolamine (DEA). Other separation methods include the application of dry adsorbents such as molecular sieves, activated carbon, iron sponge, and zinc oxide.

In the amine solvent processes, DEA solution or another amine solvent is pumped to an absorption tower where the gases are contacted and hydrogen sulfide is dissolved in the solution. The fuel gases are removed for use as fuel in process furnaces in other refinery operations. The amine–hydrogen sulfide solution is then heated and steam stripped to remove the hydrogen sulfide gas.

Current methods for removing sulfur from the hydrogen sulfide gas streams are a combination of two processes: the Claus process followed by the Beaven process, SCOT process, or the Wellman–Land process. The Claus process consists of partial combustion of the hydrogen sulfide-rich gas stream in the presence of one-third the stoichiometric quantity of air, and then reacting the resulting sulfur dioxide and unburned hydrogen sulfide in the presence of a bauxite catalyst to produce elemental sulfur.

The Claus process alone removes about 90% of the hydrogen sulfide in the gas stream. The Beaven, SCOT, or Wellman–Lord processes are used to further recover sulfur. In the Beaven process, the low levels of hydrogen sulfide in the gas stream from the Claus process can be removed by absorption in a quinone solution. The dissolved hydrogen sulfide is oxidized to form a mixture of elemental sulfur and hydroquinone. The solution is injected with air or oxygen to oxidize the hydroquinone, converting it back to quinone. The solution is then filtered or centrifuged to remove the sulfur so that the quinone can be reused.

The Beaven process is effective in removing small amounts of sulfur dioxide, carbonyl sulfide, and carbon disulfide that are not affected by the Claus process. These compounds are first converted to hydrogen sulfide at elevated temperatures in a cobalt molybdate catalyst prior to being fed to the Beaven process unit. Air emissions from sulfur recovery units consist of hydrogen sulfide, SO_x, and NO_x in the process tail gas as well as fugitive emissions and releases from vents.

In the SCOT process the sulfur compounds in the Claus tail gas are converted to hydrogen sulfide by heating and passing them through a cobalt–molybdenum catalyst with the addition of a reducing gas. The gas is then cooled and contacted with a solution of di-isopropanolamine (DIPA), which removes all but trace amounts of hydrogen sulfide. The sulfide-rich DIPA is sent to a stripper, where hydrogen sulfide gas is removed and sent to the Claus plant. The DIPA is returned to the absorption column.

Most refinery process units and equipment are manifolded into a blowdown system. Blowdown systems provide for the safe handling and disposal of liquid and gases that are either automatically vented from the process units through pressure-relief valves, or that are manually withdrawn from units. Recirculated process streams and cooling water streams are manually purged to prevent the continued buildup of contaminants in the stream. Part or all of the contents of equipment can also be purged to the blowdown system prior to shutdown before normal or emergency shutdowns.

Blowdown systems make use of an arrangement of flash drums and condensers to separate the blowdown into its vapor and liquid components. The liquid is composed of mixtures of water and hydrocarbons containing sulfides, ammonia, and other contaminants, which are sent to the wastewater treatment plant. The gaseous component typically contains hydrocarbons, hydrogen sulfide, ammonia, mercaptans, solvents, and other constituents, and is either discharged directly to the atmosphere or is combusted in a flare. The major air emissions from blowdown systems are hydrocarbons in the case of direct discharge to the atmosphere and sulfur oxides when flared.

Valves are employed in every phase of the petroleum industry where petroleum or petroleum product is transferred by piping from one point to another. There is a great variety of valve designs, but generally valves may be classified by their application as flow control or pressure relief. Manual and automatic flow control valves are used to regulate the flow of fluids through a system. Included under this classification are the gate, globe, angle, plug, and other common types of valves. These valves are subject to product leakage from the valve stem as a result of the action of vibration, heat, pressure, corrosion, or improper maintenance of valve stem packing. Pressure-relief and safety valves are used to prevent excessive pressures from developing in process vessels and lines. The relief valve designates liquid flow while the safety valve designates vapor or gas flow. These valves may develop leaks because of the corrosive action of the product or because of failure of the valve to reseat properly after blowoff. Rupture discs are sometimes used in place of pressure-relief valves. Their use is restricted to equipment in batch-type processes.

The maintenance and operational difficulties caused by the inaccessibility of many pressure-relief valves may allow leakage to become substantial. Emissions vary over a wide range. Liquid leakage results in emissions from evaporation of liquid while gas leakage results in immediate emissions. Since emissions to the atmosphere from valves are highly dependent upon maintenance, total valve losses cannot be estimated accurately. Obviously, the controlling factor in preventing leakage from valves is maintenance. An effective schedule of inspection and preventive maintenance can keep leakage at a minimum. Minor leaks that might not be detected by casual observation can be located and eliminated by thorough periodic inspections. Emissions from pressure-relief valves are sometimes controlled by manifolding to a vapor control device. Normally, these disposal systems are not designed exclusively to collect vapors from relief valves. The primary function of the system may be to collect off gases produced by a process unit, or vapors released from storage facilities, or those released by depressurizing equipment during shutdowns. Another method of control to prevent excessive emissions from relief valve leakage is the use of a dual valve with a shutoff interlock. A means of removing and repairing a detected leaking valve without waiting until the equipment can be taken out of service is thus provided. The practice of allowing a valve with a minor leak to continue in service without correction until the operating unit is shut down for general inspection is common in many refineries. This practice should be kept to a minimum.

A rupture disc is sometimes used to protect against relief valve leakage caused by excessive corrosion. The disc is installed on the pressure side of the relief valve. The space between the rupture disc and relief valve seat should be protected from pinhole leaks that could occur in the rupture disc. Otherwise, an incorrect pressure differential could keep the rupture disc from breaking at its specified pressure. This, in turn, could keep the relief valve from opening, and excessive pressures could occur in the operating equipment. One method of insuring against these small leaks in rupture discs is to install a pressure gage and a small manually operated purge valve in the system. The pressure gage would easily detect any pressure increases from even small leaks. In the event of leaks, the vessel would be removed from service, and the faulty rupture disc would then be replaced. A second, but less satisfactory method from an air pollution control standpoint is to maintain the space at atmospheric pressure by installing a small vent opening. Any minute leaks would then be vented directly to the atmosphere, and a pressure increase could not exist.

Turnarounds

In addition to the areas described above there can be significant air emissions from turnarounds. Periodic maintenance and repair of process equipment are essential to refinery operations. A major phase of the maintenance program is the shutting down and starting up of the various units, usually called a turnaround. The procedure for shutting down a unit varies from refinery to refinery and between units in a refinery. In general, shutdowns are effected by first shutting off the heat supply to the unit and circulating the feedstock through the unit as it cools. Gas oil may be blended into the feedstock to prevent its solidification as the temperature drops. The cooled liquid is then pumped out to storage facilities, leaving hydrocarbon vapors in the unit. The pressure of the hydrocarbon vapors in the unit is reduced by evacuating the contents of the various items of equipment to a disposal facility such as a fuel gas system, a vapor recovery system, or a flare or, in some cases, to the atmosphere. Discharging vapors to the atmosphere is undesirable from the standpoint of air pollution control since as much as several thousand pounds of hydrocarbons or other objectionable vapors or odors can be released during a shutdown. The residual hydrocarbons remaining in the unit after depressuring are purged out with steam, nitrogen, or water. Any purged gases should be discharged to the aforementioned disposal facilities. Condensed steam and water effluent that may be contaminated with hydrocarbons or malodorous compounds during purging should be handled by closed water-treating systems.

Tank cleaning

Storage tanks in a refinery require periodic cleaning and repair. For this purpose, the contents of a tank are removed and residual vapors are purged until the tank is considered safe for entry by maintenance crews. Purging can result in the release of hydrocarbon or odorous material in the form of vapors to the atmosphere. These vapors should be discharged to a vapor recovery system or flare. When the vapors in the tank are released to a recovery or disposal system before the tank is opened for maintenance, the emissions are considered negligible. When the stored liquid is transferred to another tank, and the emptied vessel is then opened for maintenance without purging to a recovery or disposal system, the emission to the atmosphere can be considered to be equal to the weight of hydrocarbon vapor occupying the total volume of the tank at the reported pressure.

Steam cleaning of railroad tank cars used for transporting petroleum products can similarly be a source of emissions if the injected steam and entrained hydrocarbons are vented directly to the atmosphere. Although no quantitative data are available to determine the magnitude of these emissions, the main objection to this type of operation is its nuisance-causing potential. Some measure of control of these emissions may be effected by condensing the effluent steam and vapors. The condensate can then be separated into hydrocarbon and water phases for recovery. Noncondensable vapors should be incinerated.

Use of vacuum jets

Certain refinery processes are conducted under vacuum conditions. The most practical way to create and maintain the necessary vacuum is to use steam-actuated vacuum jets, singly or in series. Barometric condensers are often used after each vacuum jet to remove steam and condensable hydrocarbons. The effluent stream from the last stage of the vacuum jet system should be controlled by condensing as much of the effluent as is practical and incinerating the noncondensables in an afterburner or heater firebox. Condensate should be handled by a closed treating system for recovery of hydrocarbons. The hot well that receives water from the barometric condensers may also have to be enclosed and any off-gases incinerated.

Compressor engine exhausts

Refining operations require the use of various types of gas compressors. These machines are often driven by internal combustion engines that exhaust air contaminants to the atmosphere. Although these engines are normally fired with natural gas and operate at essentially constant loads, some unburned fuel passes through the engine. Oxides of nitrogen are also found in the exhaust gases as a result of nitrogen fixation in the combustion cylinders.

The following are some general guidelines for best practices to reduce air emissions:

- Minimize losses from storage tanks and product transfer areas by methods such as vapor recovery systems and double seals.
- Minimize SO_x emissions either through desulfurization of fuels, to the extent feasible, or by directing the use of high-sulfur fuels to units equipped with SO_x emissions controls.
- Recover sulfur from tail gases in high-efficiency sulfur recovery units.
- Recover non-silica-based (i.e. metallic) catalysts and reduce particulate emissions.
- Use low-NO_x burners to reduce nitrogen oxide emissions.
- Avoid and limit fugitive emissions by proper process design and maintenance.
- Keep fuel usage to a minimum.
- Install vapor recovery for barge loading – although barge loading is not a factor for all refineries, it is an important emissions source for many facilities. One of the largest sources of VOC emissions identified during the Amoco/EPA study was fugitive emissions from loading of tanker barges. It was estimated that these emissions could be reduced by 98% by installing a marine vapor loss control system. Such systems could consist of vapor recovery or VOC destruction in a flare.
- Replace old boilers – older refinery boilers can be a significant source of SO_x, NO_x, and particulate emissions. It is possible to replace a large number of old boilers with a single new cogeneration plant with emissions controls.
- Eliminate use of open ponds – open ponds used to cool, settle out solids, and store process water can be a significant source of VOC emissions. Wastewater from coke cooling and coke VOC removal is occasionally cooled in open ponds, where VOCs easily escape to the atmosphere. In many cases, open ponds can be replaced with closed storage tanks.

- Remove unnecessary storage tanks from service – since storage tanks are one of the largest sources of VOC emissions, a reduction in the number of these tanks can have a significant impact. The need for certain tanks can often be eliminated through improved production planning and more continuous operations. By minimizing the number of storage tanks, tank-bottom solids and decanted wastewater may also be reduced.
- Place secondary seals on storage tanks – one of the largest sources of fugitive emissions from refineries is storage tanks containing gasoline and other volatile products. These losses can be significantly reduced by installing secondary seals on storage tanks. An Amoco/EPA joint study estimated that VOC losses from storage tanks could be reduced by 75–93%. Equipping an average tank with a secondary seal system was estimated to cost about $20,000.
- Install rupture discs and plugs – rupture discs on pressure-relief valves and plugs in open-ended valves can reduce fugitive emissions.
- Establish a leak detection and repair (LDAR) program – fugitive emissions are one of the largest sources of refinery hydrocarbon emissions. An LDAR program consists of using a portable VOC detecting instrument to detect leaks during regularly scheduled inspections of valves, flanges, and pump seals. Leaks are then repaired immediately or are scheduled for repair as quickly as possible. An LDAR program could reduce fugitive emissions by 40–64%, depending on the frequency of inspections.

6.2.2 *Wastewater*

Refineries use relatively large volumes of water, especially for cooling systems. Surface water runoff and sanitary wastewaters are also generated. The quantity of wastewaters generated and their characteristics depend on the process configuration. As a general guide, approximately 3.5–5 cubic meters (m^3) of wastewater per ton of crude are generated when cooling water is recycled. Refineries generate polluted wastewaters, containing biochemical oxygen demand (BOD) and chemical oxygen demand (COD) levels of approximately 150–250 milligrams per liter (mg/l) and 300–600 mg/l respectively, phenol levels of 20–200 mg/l, oil levels of 100–300 mg/l in desalter water and up to 5000 mg/l in tank bottoms, benzene levels of 1–100 mg/l, benzo(a)pyrene levels of less than 1 to 100 mg/l, heavy metals levels of 0.1–100 mg/l for chrome and 0.2–10 mg/l for lead, and other pollutants.

Wastewaters consist of cooling water, process water, storm water, and sanitary sewage water. A large portion of water used in petroleum refining is used for cooling. Most cooling water is recycled many times over. Cooling water typically does not come into direct contact with process oil streams and therefore contains fewer contaminants than process wastewater. However, it may contain some oil contamination due to leaks in the process equipment. It may contain significant amounts of hydrocarbons if the source of the water is groundwater that has been contaminated.

Water used in processing operations accounts for a significant portion of the total wastewater. Process wastewater arises from desalting crude oil, steam stripping, pump gland cooling, product fractionator reflux drum drains, and boiler blowdown. Because process water comes into direct contact with oil, it is usually highly contaminated. Storm water (i.e. surface water runoff) is intermittent and will contain constituents from spills, leaks in equipment and any materials that may have collected in drains. Runoff surface water also includes water coming from crude and product storage tank roof drains.

Refineries in the USA have dedicated on-site wastewater treatment facilities that discharge treated water to publicly owned treatment works (POTWs) or to surface waters under a set of National Pollution Discharge Elimination System (NPDES) permits. Treatment works are generally comprised of primary and secondary treatment. Primary wastewater treatment consists of the separation of oil, water, and solids in two stages; during the first stage, an API separator, a corrugated plate interceptor, or other separator design is used. Wastewater slowly moves through the separator allowing free oil to float to the surface where it is skimmed off, and solids that settle to the bottom are scraped away to a sludge-collecting hopper. The second stage utilizes physical or chemical methods to separate emulsified oils from the wastewater. Physical methods may include the use of a series of settling ponds with long retention times, or dissolved air flotation (DAF) units are employed. In DAF, air is bubbled through the wastewater, and both oil and suspended solids are skimmed off the top. Chemicals, such as ferric hydroxide or aluminum hydroxide, are used to coagulate impurities into a froth or sludge that can be more easily skimmed off the top. Some wastes associated with the primary treatment of wastewater may be considered hazardous and include API separator sludge, primary treatment sludge, sludges from other gravitational separation techniques, float from DAF units, and wastes from settling ponds.

In secondary treatment, dissolved oil and other organic pollutants may be consumed biologically by microorganisms. Biological treatment may require the addition of oxygen through a number of different techniques, including activated sludge units, trickling filters, and rotating biological contactors. Secondary treatment generates biomass waste that is typically treated anaerobically and then dewatered.

Some refineries employ an additional stage of wastewater treatment called polishing to meet discharge limits. The polishing step can involve the use of activated carbon, anthracite coal, or sand to filter out any remaining impurities, such as biomass, silt, trace metals and other inorganic chemicals, as well as any remaining organic chemicals.

Some refinery wastewater streams are treated separately, prior to the wastewater treatment plant, to remove contaminants that would not easily be treated after mixing with other wastewater. One example is the sour water drained from distillation reflux drums. Sour water contains dissolved hydrogen sulfide and other organic sulfur compounds and ammonia, which are stripped in a tower with gas or steam before being discharged to the wastewater treatment plant.

For those readers who are not familiar with some of the practices and operations of refinery wastewater treatment plants, Figures 6.1–6.5 provide some examples.

Figure 6.1 Oil–water separator.

Wastewater treatment plants are significant sources of air emissions and solid wastes. Air releases arise from fugitive emissions from the numerous tanks, ponds, and sewer system drains. Solid wastes are generated as sludges from the various treatment units.

Many older refineries had poor housekeeping practices, poor preventive maintenance programs, and leaking storage tanks. These have resulted in unintentional releases of liquid hydrocarbons to groundwater and surface waters. At some refineries contaminated groundwater has migrated off-site and resulted in continuous "seeps" to surface waters. An example of this is the old Gulf refinery in Hooven, Ohio. The reader can find a considerable amount of information on this site from a search on the Web. This refinery was built in the early 1930s and was purchased by Chevron in 1984 and subsequently shut down in 1986. Official statements by Chevron report 5 million gallons of gasoline constituents have contaminated the groundwater aquifer beneath the city. An examination of Chevron documents in a litigation show that the actual level of groundwater contamination may be as high as 26 million gallons, with a record of decision (ROD) allowing at least 50 years of remediation and monitoring. Chevron unpublished studies show that it could take as long as 500 years for the groundwater to be restored to maximum allowable concentrations for some constituents.

Figure 6.2 Flotation units.

Oil–water effluent systems are found in the three phases of the petroleum industry, namely production, refining, and marketing. The systems vary in size and complexity though their basic function remains the same, i.e. to collect and separate wastes, to recover valuable oils, and to remove undesirable contaminants before discharge. In the production of crude oil, wastes such as oily brine, drilling muds, tank bottoms, and free oil are generated. Of these, the oilfield brines present the most difficult disposal problem because of the large volume encountered. Community disposal facilities capable of processing the brines to meet local water pollution standards are often set up to handle the treatment of brines. Among the traditional methods of disposal of brines has been injection into underground formations. A typical collection system associated with the crude-oil production phase of the industry usually includes a number of small gathering lines or channels transmitting wastewater from wash tanks, leaky equipment, and treaters to an earthen pit, a concrete-lined sump, or a steel wastewater tank. A pump decants wastewater from these containers to water-treating facilities before injection into underground formations or disposal to sewer systems. Any oil accumulating on the surface of the water is skimmed off to storage tanks.

The effluent disposal systems found in refineries are larger and more elaborate than those in the production phase. A typical modern refinery gathering system

Figure 6.3 Clarifier.

Figure 6.4 Small retention pond for solids settling.

Figure 6.5 Aeration pond.

usually includes gathering lines, drain seals, junction boxes, and channels of vitrified clay or concrete for transmitting wastewater from processing units to large basins or ponds used as oil–water separators. These basins are sized to receive all effluent water, sometimes even including rain runoff, and may be earthen pits, concrete-lined basins, or steel tanks. Liquid wastes discharging to these systems originate at a wide variety of sources such as pump glands, accumulators, spills, cleanouts, sampling lines, relief valves, and many others. The types of liquid wastes may be classified as wastewater with: oil present as free oil or emulsified oil, or as oil coating on suspended matter; or chemicals present as suspensoids, emulsoids, or solutes. These chemicals include acids, alkalis, phenols, sulfur compounds, clay, and others. Emissions from these varied liquid wastes can best be controlled by properly maintaining, isolating, and treating the wastes at their source, by using efficient oil–water separators, and by minimizing the formation of emulsions.

The wastewater from the process facilities and treating units just discussed flows to the oil–water separator for recovery of free oil and settleable solids. Factors affecting the efficiency of separation include temperature of water, particle size, density, and amounts and characteristics of suspended matter. Stable emulsions are not affected by gravity-type separators and must be treated separately. The oil–water separator design must provide for efficient inlet and outlet construction, sediment collection mechanisms, and oil skimmers. Reinforced concrete construction has been found most desirable for reasons of economy, maintenance, and efficiency.

The effluent water from the oil–water separator may require further treatment before final discharge to municipal sewer systems, channels, rivers, or streams. The type and extent of treatment depend upon the nature of the contaminants present, and on the local water pollution ordinances governing the concentration and amounts of contaminants to be discharged in refinery effluent waters. The methods of final-effluent clarification to be briefly discussed here include (1) filtration, (2) chemical flocculation, and (3) biological treatment. Several different types of filters may be used to clarify the separator effluent. Hay-type filters, sand filters, and vacuum precoat filters are the most common. The selection of any one type depends upon the properties of the effluent stream and upon economic considerations. Methods of treatment are either by sedimentation or flotation. In sedimentation processes, chemicals such as copper sulfate, activated silica, alum, and lime are added to the wastewater stream before it is fed to the clarifiers. The chemicals cause the suspended particles to agglomerate and settle out. Sediment is removed from the bottom of the clarifiers by mechanical scrapers. Effectiveness of the sedimentation techniques in the treatment of separator effluents is limited by the small oil particles contained in the wastewater. These particles, being lighter than water, do not settle out easily. They may also become attached to particles of suspended solids and thereby increase in buoyancy. In the flotation process a colloidal floc and air under pressure are injected into the wastewater. The stream is then fed to a clarifier through a backpressure valve that reduces the pressure to atmospheric. The dissolved air is suddenly released in the form of tiny bubbles that carry the particles of oil and coalesced solids to the surface, where they are skimmed off by mechanical flight scrapers. Of the two, the flotation process has the potential to become the more efficient and economical. Biological treating units such as trickling filters, activated sludge basins, and stabilization basins have been incorporated into modern refinery waste disposal systems. By combining adsorption and oxidation, these units are capable of reducing oil, biological oxygen demand, and phenolic content from effluent water streams. To prevent the release of air pollutants to the atmosphere, certain pieces of equipment, such as clarifiers, digesters, and filters used in biological treatment should be covered and their contents vented to recovery facilities or incinerated.

From an air pollution standpoint the most objectionable contaminants emitted from liquid waste streams are hydrocarbons, sulfur compounds, and other malodorous materials. The effect of hydrocarbons in smog-producing reactions is well known, and sulfur compounds such as mercaptans and sulfides produce very objectionable odors, even in high dilution. These contaminants can escape to the atmosphere from openings in the sewer system, open channels, open vessels, and open oil–water separators. The large exposed surface area of these separators requires that effective means of control be instituted to minimize hydrocarbon losses to the atmosphere from this source. The most effective means of control of hydrocarbon emissions from oil–water separators has been the covering of forebays or primary separator sections. Either fixed roofs or floating roofs are considered acceptable covers. Separation and skimming of over 80% of the floatable oil layer is effected in the covered sections. Thus, only a minimum of oil is contained in the effluent water, which flows under concrete curtains to the open afterbays or secondary separator sections. The explosion hazard associated with fixed roofs is not present in a floating-roof installation. These roofs are similar to those developed for storage tanks. The floating covers are built to fit into bays with about 1 inch of clearance around the perimeter. Fabric or rubber may be used to seal the gap between the roof edge and the container wall. The roofs are fitted with access manholes, skimmers, gage hatches, and supporting legs. In operation, skimmed oil flows through lines from the skimmers to a covered tank (floating roof or connected to vapor recovery) or sump and then is pumped to de-emulsifying processing facilities. Effluent water from the oil–water separator is handled in the manner described previously.

In addition to covering the separator, open sewer lines that may carry volatile products are converted to closed, underground lines with waterseal-type vents. The contents of junction boxes are vented to vapor recovery facilities, and steam is used to blanket the sewer lines to inhibit formation of explosive mixtures. Accurate calculation of the hydrocarbon losses from separators fitted with fixed roofs is difficult because of the many variables of weather and refinery operations involved.

Isolation of certain odor- and chemical-bearing liquid wastes at their source for treatment before discharge of the water to the refinery wastewater-gathering system has been found to be the most effective and economical means of minimizing odor and chemicals problems. The unit that is the source of wastes must be studied for possible changes in the operating process to reduce wastes. In some cases the wastes from one process may be used to treat the wastes from another. Among the principal streams that are treated separately are oil-in-water emulsions, sulfur-bearing waters, acid sludge, and spent caustic wastes.

Oil-in-water emulsions are wastes that can be treated at their source. An oil-in-water emulsion is a suspension of oil particles in water that cannot be divided effectively by means of gravity alone. Gravity-type oil–water separators are generally ineffective in breaking the emulsions, and means are provided for separate treatment where the problem is serious. Oil-in-water emulsions are objectionable in the drainage system since the separation of otherwise recoverable oil may be impaired by their presence. Moreover, when emulsions of this type are discharged into large bodies of water, the oil is released by the effect of dilution, and serious pollution of the water may result. Formation of emulsions may be minimized by proper design of process equipment and piping. Both physical and chemical methods are available for use in breaking emulsions. Physical methods of separation include direct application of heat, distillation, centrifuging, filtration, and use of an electric field. The effectiveness of any one method depends upon the type of emulsion to be treated.

Sulfur-bearing waters

Sulfides and mercaptans are removed from wastewater streams by various methods. Some refineries strip the wastewater in a column with live steam. The overhead vapors from the column are condensed and collected in an accumulator from which the noncondensables flow to sulfur recovery facilities or are incinerated. Flue gas has also been used as the stripping medium. Bottoms water from steam stripping towers, being essentially sulfide free, can then be drained to the refinery's sewer system. Oxidation of sulfides in wastewater is also an effective means of treatment. Air and heat are used to convert sulfides and mercaptans to thiosulfates, which are water soluble and not objectionable. Chlorine is also used as an oxidizing agent for sulfides. It is added in stoichiometric quantities proportional to the wastewater. This method is limited by the high cost of chlorine. Water containing dissolved sulfur dioxide has been used to reduce sulfide concentration in wastewaters. For removing small amounts of hydrogen sulfide, copper sulfate and zinc chloride have been used to react with and precipitate the sulfur as copper and zinc sulfides. Hydrogen sulfide may be released, however, only if the water treated with these compounds contacts an acid stream.

Acid sludge

The acid sludge produced from treating operations varies with the stock treated and the conditions of treatment. The sludge may vary from a low-viscosity liquid to a solid. Methods of disposal of this sludge are many and varied. Basically, they may be considered under the following general headings: disposal by burning as fuel, or dumping in the ground; processing to produce by-products such as ammonium sulfate, metallic sulfates, oils, tars, and other materials; processing for recovery of acid. The burning of sludge results in discharge to the atmosphere of excessive amounts of sulfur dioxide and sulfur trioxide from furnace stacks. If sludge is solid or semi-solid it may be buried in specially constructed pits. This method of disposal, however, creates the problem of acid leaching out to adjacent waters. Recovery of sulfuric acid from sludge is accomplished essentially by either hydrolysis or thermal decomposition processes. Sulfuric acid sludge is hydrolyzed by heating it with live steam in the presence of water. The resulting product separates into two distinct phases. One phase consists of dilute sulfuric acid with a small amount of suspended carbonaceous material, and the second phase, of a viscous acid–oil layer. The dilute sulfuric acid may be (1) neutralized by alkaline wastes, (2) reacted chemically with ammonia–water solution to produce ammonium sulfate for fertilizer, or (3) concentrated by heating.

6.2.3 Waste solids

Refineries also generate solid wastes and sludges (ranging from 3 to 5 kg per ton of crude processed), 80% of which may be considered hazardous because of the presence of toxic organics and heavy metals. Accidental discharges of large quantities of pollutants can occur as a result of abnormal operation in a refinery and potentially pose a major local environmental hazard.

Solid wastes are generated from the refining processes, and petroleum handling operations, as well as wastewater treatment. Both hazardous and non-hazardous wastes are generated. Refinery wastes are typically in the form of sludges (including sludges from wastewater treatment), spent process catalysts, filter clay, and incinerator ash. Treatment of these wastes includes incineration, land treating off-site, land filling on-site, land filling off-site, chemical fixation, neutralization, and other treatment methods.

A significant portion of the non-petroleum product outputs of refineries is transported off-site and sold as by-products. These outputs include sulfur, acetic acid, phosphoric acid, and recovered metals. Metals from catalysts and from the crude oil that have deposited on the catalyst during the production often are recovered by third-party recovery facilities.

Storage tanks are used to store crude oil and intermediate process feeds for cooling and further processing. Finished petroleum products are also kept in storage tanks before transport off-site. Storage tank bottoms are mixtures of iron rust from corrosion, sand, water, and emulsified oil and wax, which accumulate at the bottom of tanks. Liquid tank bottoms (primarily water and oil emulsions) are periodically drawn off to prevent their continued buildup. Tank-bottom liquids and sludge are also removed during periodic cleaning of tanks for inspection.

The following are best practices aimed at reducing the loadings to wastewater treatment works and thus reducing volumes of sludge that require containment and costly disposal:

- Segregate process waste streams – a significant portion of refinery waste arises from oily sludges found in combined process/storm sewers. Segregation of the relatively clean rainwater runoff from the process streams can reduce the quantity of oily sludges generated. Furthermore, there is a much higher potential for recovery of oil from smaller, more concentrated process streams.
- Control solids entering sewers – solids released to the wastewater sewer system can account for a large portion of a refinery's oily sludges and may overload the wastewater treatment plant. Solids entering the sewer system (primarily soil particles) become coated with oil and are deposited as oily sludges in the API oil–water separator. Because a typical sludge has a solids content of 5–30% by weight, preventing 1 pound of solids from entering the sewer system can eliminate 3–20 pounds of oily sludge. In a study at the Jordan oil refinery led by one of the authors, we found nearly 1200 tons of solids per year entering the refinery sewer system. Methods used to control solids include: using a street sweeper on paved areas, paving unpaved areas, planting ground cover on unpaved areas, relining sewers, cleaning solids from ditches and catch basins, and reducing heat exchanger bundle cleaning solids by using antifoulants in cooling water.
- Identify benzene sources and install upstream water treatment – benzene in wastewater can often be treated more easily and effectively at the point it is generated rather than at the wastewater treatment plant after it is mixed with other wastewater.

- Train personnel to reduce solids in sewers – a facility training program that emphasizes the importance of keeping solids out of the sewer systems will help reduce that portion of wastewater treatment plant sludge arising from the everyday activities of refinery personnel.
- Train personnel to prevent soil contamination – contaminated soil can be reduced by educating personnel on how to avoid leaks and spills.
- Reduce the use of 55-gallon drums – replacing 55-gallon drums with bulk storage can minimize the chances of leaks and spills that contaminate groundwater, enter into stormwater runoff, and may enter into the plant sewer system.
- Refurbish or eliminate underground piping – underground piping can be a source of undetected releases to the soil and groundwater. Inspecting, repairing, or replacing underground piping with surface piping can reduce or eliminate these potential sources.
- Minimize solids leaving the desalter – solids entering the crude distillation unit are likely to eventually attract more oil and produce additional emulsions and sludges. The amount of solids removed from the desalting unit should therefore be maximized. A number of techniques can be used, such as: using low-shear mixing devices to mix desalter wash water and crude oil; using lower pressure water in the desalter to avoid turbulence; and replacing the water jets used in some refineries with mud rakes, which add less turbulence when removing settled solids.
- Minimize cooling tower blowdown – the dissolved solids concentration in the recirculating cooling water is controlled by purging or blowing down a portion of the cooling water stream to the wastewater treatment system. Solids in the blowdown eventually create additional sludge in the wastewater treatment plant. However, the amount of cooling tower blowdown can be lowered by minimizing the dissolved solids content of the cooling water. A significant portion of the total dissolved solids in the cooling water can originate in the cooling water makeup stream in the form of naturally occurring calcium carbonates. Such solids can be controlled either by selecting a source of cooling tower makeup water with less dissolved solids or by removing the dissolved solids from the makeup water stream. Common treatment methods include cold lime softening, reverse osmosis, or electrodialysis.
- Control heat exchanger cleaning solids – in many refineries, using high-pressure water to clean heat exchanger bundles generates and releases water and entrained solids to the refinery wastewater treatment system. Exchanger solids may then attract oil as they move through the sewer system and may also produce finer solids and stabilized emulsions that are more difficult to remove. Solids can be removed at the heat exchanger cleaning pad by installing concrete overflow weirs around the surface drains or by covering drains with a screen. Other ways to reduce solids generation are by using antifoulants on the heat exchanger bundles to prevent scaling and by cleaning with reusable cleaning chemicals that also allow for the easy removal of oil.
- Control surfactants in wastewater – surfactants entering the refinery wastewater streams will increase the amount of emulsions and sludges generated. Surfactants can enter the system from a number of sources, including: washing unit pads with detergents; treating gasolines with an end point over 400°F, thereby producing spent caustics; cleaning tank truck interiors; and using soaps and cleaners for miscellaneous tasks. In addition, the overuse and mixing of the organic polymers used to separate oil, water, and solids in the wastewater treatment plant can actually stabilize emulsions. The use of surfactants should be minimized by educating operators, routing surfactant sources to a point downstream of the DAF

unit, and by using dry cleaning, high-pressure water or steam to clean oil surfaces of oil and dirt.

- Recycle and regenerate spent caustics – caustics used to absorb and remove hydrogen sulfide and phenol contaminants from intermediate and final product streams can often be recycled. Spent caustics may be saleable to chemical recovery companies if concentrations of phenol or hydrogen sulfide are high enough. Process changes in the refinery may be needed to raise the concentration of phenols in the caustic to make recovery of the contaminants economical. Caustics containing phenols can also be recycled on-site by reducing the pH of the caustic until the phenols become insoluble, thereby allowing physical separation. The caustic can then be treated in the refinery wastewater system.
- Use oily sludges as feedstock – many oily sludges can be sent to a coking unit or the crude distillation unit, where it becomes part of the refinery products. Sludge sent to the coker can be injected into the coke drum with the quench water, injected directly into the delayed coker, or injected into the coker blowdown contactor used in separating the quenching products. Use of sludge as a feedstock has increased significantly in recent years and is currently carried out by most refineries. The quantity of sludge that can be sent to the coker is restricted by coke quality specifications, which may limit the amount of sludge solids in the coke. Coking operations can be upgraded, however, to increase the amount of sludge that they can handle.
- Control and reuse the fluidized bed catalytic cracking unit (FCCU) and coke fines – significant quantities of catalyst fines are often present around the FCCU catalyst hoppers and reactor and regeneration vessels. Coke fines are often present around the coker unit and coke storage areas. The fines can be collected and recycled before being washed to the sewers or migrating off-site via the wind. Collection techniques include dry sweeping the catalyst and coke fines and sending the solids to be recycled or disposed of as non-hazardous waste. Coke fines can also be recycled for fuel use. Another collection technique involves the use of vacuum ducts in dusty areas (and vacuum hoses for manual collection), which run to a small baghouse for collection.
- Recycle lab samples – lab samples can be recycled to the oil recovery system.
- Improve recovery of oils from oily sludges – because oily sludges make up a large portion of refinery solid wastes, any improvement in the recovery of oil from the sludges can significantly reduce the volume of waste. There are a number of technologies currently in use to mechanically separate oil, water, and solids, including belt filter presses, recessed chamber pressure filters, rotary vacuum filters, scroll centrifuges, disc centrifuges, shakers, thermal driers, and centrifuge–drier combinations.
- Reduce the generation of tank bottoms – tank bottoms from crude oil storage tanks constitute a large percentage of refinery solid waste and pose a particularly difficult disposal problem due to the presence of heavy metals. Tank bottoms are comprised of heavy hydrocarbons, solids, water, rust, and scale. Minimization of tank bottoms is carried out most cost-effectively through careful separation of the oil and water remaining in the tank bottom. Filters and centrifuges can also be used to recover the oil for recycling.
- Regenerate or eliminate filtration clay – clay from refinery filters must periodically be replaced. Spent clay often contains significant amounts of entrained hydrocarbons and therefore must be designated as hazardous waste. Backwashing spent clay with

water or steam can reduce the hydrocarbon content to levels such that the clay can be reused or handled as a non-hazardous waste. Another method used to regenerate clay is to wash the clay with naphtha, dry it by steam heating, and then feed it to a burning kiln for regeneration. In some cases clay filtration can be replaced entirely with hydrotreating.

- Minimize FCCU decant oil sludge – decant oil sludge from the FCCU can contain significant concentrations of catalyst fines. These fines often prevent the use of decant oil as a feedstock or require treatment that generates an oily catalyst sludge. Catalysts in the decant oil can be minimized by using a decant oil catalyst removal system. One system incorporates high-voltage electric fields to polarize and capture catalyst particles in the oil. The amount of catalyst fines reaching the decant oil can be minimized by installing high-efficiency cyclones in the reactor to shift catalyst fines losses from the decant oil to the regenerator, where they can be collected in the electrostatic precipitator.
- Use non-hazardous degreasers – spent conventional degreaser solvents can be reduced or eliminated through substitution with less toxic and/or biodegradable products.
- Eliminate chromates as an anticorrosive – chromate-containing wastes can be reduced or eliminated in cooling tower and heat exchanger sludges by replacing chromates with less toxic alternatives such as phosphates.
- Use high-quality catalysts – by using catalysts of a higher quality, process efficiencies can be increased while the required frequency of catalyst replacement can be reduced.
- Replace ceramic catalyst supports with activated alumina supports – activated alumina supports can be recycled with spent alumina catalyst.
- Thermal treatment of applicable sludges – the toxicity and volume of some deoiled and dewatered sludges can be further reduced through thermal treatment. Thermal sludge treatment units use heat to vaporize the water and volatile components in the feed, and leave behind a dry solid residue. The vapors are condensed for separation into the hydrocarbon and water components. Noncondensable vapors are either flared or sent to the refinery amine unit for treatment and use as refinery fuel gas.
- Implement and enforce good housekeeping practices – good housekeeping practices prevent waste by better handling of both inputs and wastes without making significant modifications to current production technology. If inputs are handled better, they are less likely to become wastes inadvertently through spills or outdating. If wastes are handled better, they can be managed in the most cost-effective manner. By way of examples and to further emphasize practices to avoid, examine the photographs in Figures 6.6–6.14.

6.3   Other best management practices

The following is a list of other best management practices that can lead to improved environmental performance:

- Implement SHE (Safety, Health, and Environment) audits combined with a data management system for scheduling and following up on corrective actions.
- Dike areas and collect spillage during barrel unloading and recycle products.
- Install automatic tank cleaning and oil recovery systems.

Figure 6.6 Example of poor preventive maintenance. Leaks lead to fugitive air emissions and additional loadings to the wastewater treatment plant.

Figure 6.7 Allowing steam traps to malfunction not only results in losses in water and energy credits, but may add to fugitive air emissions. The condensate also washes solids into the plant sewer, thus adding additional loadings to the wastewater treatment plant.

- Develop written spill prevention and countermeasures programs and train employees.
- Establish a waste minimization program with appropriate metrics for monitoring environmental performance.
- Establish an environmental action plan and an environmental management plan.
- Devise and implement a drum inspection program.
- Label and inventory all waste drums and establish a policy for regular removal and disposal.
- Establish written guidelines for recycling and monitoring and enforce them.
- Survey technologies for sludge and contaminated soil remediation.
- Provide 40-hour hazard materials handling training or equivalent.
- Provide right-to-know training.
- Prepare written health and safety plans (HASPs) and reinforce with frequent training.
- Critically review and revise emergency response programs and policies on a yearly basis.
- Establish chemical inventory and restricted use policy.
- Develop a drum management program.
- Store chemicals in secured areas that are diked.
- Implement a stormwater management program.

Figure 6.8 Refineries have thousands of piping connectors such as flanges, couplings, and valves, all of which leak to some degree. These are sources of fugitive emissions. Leak detection and repair programs are necessary to minimize these emissions and product losses.

- When exchangers become fouled, solids require removal. Steps to minimize solids washing to sewers include:
  - Installation of concrete overflow weirs around exchanger pads and drains to retain solids from tube bundle cleaning.
  - Install temporary covers over sewer drains during cleaning operation.
  - Restrict cleaning to designated areas that are diked or contained for solids collection.
- To prevent solids from entering the plant sewers and overloading the wastewater treatment plant:
  - Use street sweepers on paved areas to remove trash and debris before it can be washed into sewers.
  - Install paving and plant ground cover in unpaved areas near sewers.
  - Increase inspection and repairs of sewer line breaks and lining sewers.
  - Implement a practice of periodic cleaning of solids from ditches and catch basins and use vacuuming of solids.
  - Use cyclone separators upstream of the API separator to reduce fines contributing to sludge formation.
  - Collect (sweep and shovel) FCCU catalyst spills.

Figure 6.9 Part of an asphalt plant where lack of preventive maintenance has allowed the equipment to deteriorate to the point of excessive emissions.

- Use filters at sewer drains in coking units to keep coke fines out of oily water sewers.
- Use hydroclones to recover fines that escape into the sewer.
- Perform periodic inspection and repair of above- and underground piping.
- Install tank overfill prevention systems.
- Install pavement in place of bare ground or other surfaces under major pipe racks to facilitate leak detection.
- Use detectors to reduce oil drainage during tank draws, and use automatic water draws on product and crude oil tanks.
- Segregation and flow reduction of stormwater and wastewater to separate stormwater and oily water sewers to reduce wastewater flows to the treatment plant, contamination of stormwater with hydrocarbons, and sludge formation:
  - Dikes can be installed in selected process areas to prevent drainage of hydrocarbon-bearing streams into stormwater sewers.
  - Impound stormwater from areas of potential contamination (e.g. tank farms) for sampling to verify whether treatment is necessary (e.g. the so-called first flush runoff from areas that may be somewhat oil contaminated but that are unlikely to produce contaminated runoff after a certain initial amount of rain has fallen).
  - Use collected rainwater as wash water for process use to minimize runoff flow rates, although the potential is largely limited to clean stormwater runoff that does not contain entrained soils and sand.

Figure 6.10 Good housekeeping practices should be performed safely. This worker is steam cleaning an area where gasoline and oil spills have occurred. Because he is not wearing a respirator he places himself at risk from exposure to volatile hydrocarbons.

- Divert waste streams with primarily inorganic contaminants (e.g. streams such as stripped sour water or boiler blowdown) directly to biological treatment downstream of the API separator and dissolved air flotation (DAF) units to minimize sludge formation in these units.
- Reuse and recycle wash water to the maximum extent possible. Applications as desalter feed water and wash water for further unit and tank washing are examples. Water can be injected into the crude and vacuum distillation unit overhead streams for corrosion control, and condensed stripping steam are often suitable as desalter makeup water. Stripped sour water is also a good source of makeup water for the desalter.
- Survey oily water and stormwater sewers with cameras and dyes to detect cross-connections between the two systems. Eliminating cross-connections will reduce stormwater intrusion into the oily water system and reduce the amount of hazardous waste generated in the oily water system.
- Minimize sample losses to the sewer system:
  - Implement closed-loop sampling systems so that sample streams return to the process and are not sent to the sewer. Such systems were originally installed for benzene-containing streams due to benzene NESHAPS rules. Closed-loop systems can easily be installed to channel flow from a pump discharge line to a suction line on the same pump or to channel flow around a control valve.

Figure 6.11 These spills enter into the stormwater and are carried both off-site and into the plant sewer, where they contribute to toxic sludge generation.

- Recycle laboratory samples of crude oils and samples of refined and intermediate product streams to their oil recovery systems after the laboratory has finished its analyses.
- Implement multiple practices to reduce the amount of soil, sand, and trash entering the sewer systems:
  - Use street sweepers on paved areas to remove trash before it can be washed into sewers.
  - Use beds of small rock installed on earthen tank farm floors to impede entrainment of soil and sand in rainwater that falls on the tank farm areas.
  - Install curbs and berms to protect some sewer drains from solids in stormwater runoff and wash water.
  - Use erosion control pipe trenches and catch basins.
  - Eliminate the use of sandbags or burlap bags topped with sand as covers to plug sewers during maintenance to avoid potential deterioration of the sandbags and spillage of sand into sewers.
  - Replace sandbags and burlap bags with temporary seals, lead blankets, or other commercial devices.
  - Implement methods of controlling and containing sandblast grit (which contains metal, old paint, and primer, some of which may contain lead) to keep it out of the sewer system.

Figure 6.12 Replacing 55-gallon drums with bulk storage can minimize the chances of leaks and spills that contaminate groundwater, enter into stormwater runoff, and may enter into the plant sewer system.

- Practice segregating toxic sandblast media as well as segregate sand by the type of paint it is used to remove (i.e. leaded and non-leaded).
- Use absorbents (e.g. diatomaceous earth, vermiculite) rather than sand for cleaning up oily surfaces. Absorbents are easier to remove than sand and require less water wash for final cleanup.
- Use cyclonic separators upstream of the API gravity separators to reduce the quantity of fines contributing to sludge formation in the separators.
- Minimize spent catalyst waste (note spent catalyst waste disposal represents a major cost and potentially large savings to the extent that catalysts can be recycled and their useful life extended). Spent catalyst disposal is largely a solid waste and dangerous waste issue, but catalysts also impact the wastewater systems of refineries in several ways. First, the regeneration of catalytic reformer catalyst can produce dioxins and furans as unintended by-products that can reach the sewer system. Second, FCCU catalyst fines can pose a solids control problem and can be washed into the sewer system. Third, change-out of catalysts can release dust and fines that eventually wash into the sewer system. Among the pollution prevention practices to consider are:
    - Optimize operating parameters affecting catalyst life in all major processing units, to provide better removal of catalyst poisons from feed streams, and to upgrade feedstock quality where feasible to extend catalyst life.

Figure 6.13 Small spills in themselves are not necessarily a problem. But cumulatively, spills add up to many hundreds to thousands of events over the course of a year and represent both harmful VOC emissions to the atmosphere and community exposure, plus millions of dollars in lost product to the refinery.

- Improve hydrocarbon recovery from spent sulfuric acid in alkylation units (e.g. by contacting alkylate product with primary settler acid discharge so that heavy hydrocarbons in the product absorb light hydrocarbons in the spent acid). Improved hydrocarbon recovery would decrease sewer losses and reduce sludge formation.
- Consider agricultural use of spent polymerization unit catalyst, which is composed of phosphoric acid on a silica–alumina base. The phosphoric acid is a good source of phosphorus for cultivated plants.
- Identify alternative disposal options for alkylation unit sludge:
  - Use as a fluxing substitute in metal refining and as a raw material for manufacturing hydrofluoric acid.
  - Sludge generation can be decreased in some units by replacing insoluble neutralizing agents (e.g. lime) with soluble agents (e.g. sodium hydroxide). This approach does have the disadvantage of increasing fluoride levels in the wastewater and refinery outfall. Some refineries neutralize with agents that precipitate fluoride in the form of a marketable by-product (e.g. as calcium fluoride).
  - Sludge generation has been decreased in some refineries by sending acid regenerator bottoms to other processing units rather than to the neutralization pit, where the sludge forms.

Figure 6.14 Loading liquid products without purging systems or bottom-filling practices results in sources of fugitive emissions.

- Minimize amine losses and sludge generation in amine units – note: amine treating units are used to remove hydrogen sulfide (H_2S) from different refinery sour gas streams, producing a low-sulfur fuel gas and, after regeneration of the amine in a stripper, an acid gas stream containing the H_2S that is sent to the sulfur recovery unit. The main solvents involved in amine systems in refineries are monoethanolamine (MEA), diethanolamine (DEA), diglycolamine (DGA), di-isopropanolamine (DIPA), methyldiethanolamine (MDEA), and various proprietary formulations of these amines and additives. Selection of the amine for a given application is typically a function of selectivity of absorption to H_2S and CO_2. A portion of the recovered amine stream from the regenerator is blown down to the sewer system to prevent buildup of impurities. The amines in this blowdown stream can interfere with performance of biological organisms in the wastewater treatment plant. Refiners have addressed several options to reduce amine losses as well as to minimize sludge generation with the following practices:
  - Employ a sump to retain amines drained from sludge filters in the Claus/tail gas unit during filter bag change-outs. These amines would otherwise be lost to the wastewater treatment unit.
  - Replace cloth filters with metal filters for sludge filtration to reduce maintenance and eliminate amine discharges associated with filter change-outs.
  - Replace MEA with MDEA to reduce formation of heat-stable salts and minimize quantities of amine sludge and spent amine solution from tail gas units.

- Use additives to minimize heat-stable salts in MDEA systems. MEA has had widespread use. It is inexpensive and highly reactive. However, it is irreversibly degraded by impurities. MDEA has the advantage of a high selectivity to H_2S but not to CO_2.
- Spent caustic recycle – caustic treating is used throughout a refinery to remove hydrogen sulfide and phenolic compounds from various streams. Spent caustic streams are generally treated in the wastewater treatment facilities. Various options for recycling and minimizing spent caustic are:
  - Cascading of caustic streams from one unit to another provides an opportunity to optimize caustic use while reducing the quantity of fresh caustic needed as well as the total wastewater treatment load. Some specialty chemical companies will buy spent caustic streams from refiners to recover the phenol value, although the cost-effectiveness of this approach depends on several factors, including proximity of the recovery facilities to the refinery.
  - Some refiners have installed commercial caustic regeneration units.
- Use oily sludge as coker feedstock – refineries with coker operations can in many cases use relatively small quantities of waste and residual streams as coker feedstock without affecting petroleum coke product quality. Oil-containing sludge is an example of a potential coker feedstock that would otherwise have to be disposed of as a hazardous waste or fed to a process (such as a filter press) to recover the oil. (Sludge sources that have been successfully fed to a coker unit include exchanger bundle sludge, filter cake from tank cleaning, primary treatment sludge, oil emulsions and slop oil emulsion solids, laboratory wastes, etc.). Coke product specifications are typically the limiting factor in determining how much of this material can be processed.
- Desalter improvements – desalter operations are a significant source of contaminated wastewater. In addition to implementing operating and maintenance improvements, several refiners have evaluated desalter modification or replacement. Successful modifications of desalter internals have been made to improve efficiency, including replacement of internals with more efficient electrical equipment to improve the ability to coalesce water droplets in the emulsion, thereby improving oil–water separation.
  - Refiners have evaluated methods of elimination of desalters involving replacing them with other processes, including dehydration of oil with emulsion breakers.
  - Evaluate the use of various processing steps to treat desalter water before it enters the sewer system to recover remaining oil and reduce waste loads. Centrifugation and air flotation are potential steps to reduce sewer loads.
- Alternative catalysts for HF alkylation units – alkylation catalysts are one of two strong acids, hydrofluoric acid (HF) or sulfuric acid (H_2SO_4). In both of these systems, acid is added continuously as a liquid. Care must be taken not to allow these acids to reach the wastewater treatment system. In sulfuric acid units, spent acid is recycled to produce fresh sulfuric acid. The HF units use less acid per volume of alkylate produced, and the HF acid is consumed by feed contaminants. Thus, HF units do not recycle the acid as do sulfuric acid units.
  - New processes have been evaluated that would employ solid acid catalysts and small quantities of liquid acid catalysts to replace HF and H_2SO_4, thereby eliminating the acid-soluble oil stream, the neutralization of which generates sludge.
  - Use of solid acid catalyst can reduce quantities of adsorbents (such as molecular sieves, alumina, sand, and salt) used and the quantity of spent adsorbents to be disposed of as hazardous waste.

- Implement a facility-wide water reuse evaluation – overall water reuse evaluations within refineries are based on influent water purchase and treatment costs, wastewater treatment costs, permit limitations, and various non-economic factors (e.g. community relations). The objective of a program is to identify all of the major influent streams to and effluent streams from each process and utility unit as well as from the waste treatment area. Each stream is then characterized in terms of pollutants, composition, flow characteristics, and other parameters. Matching influent requirements to effluent parameters will identify effluent streams that are potential candidates for reuse as influent streams to other units (e.g. stripped sour water, an effluent stream, as a candidate for desalter feed water, an influent stream). A potential candidate would be an effluent stream that already matches influent requirements in one or more other units or one that with relatively minor treatment steps would match influent requirements. Such steps might include treatment with biocides, pH adjustment, filtration, or other procedures that generally do not entail high capital or operating costs.
- Minimize cooling tower blowdown rates and pollutants:
  - To reduce the dissolved solids level in the cooling tower makeup water, apply water softening, reverse osmosis, or electrodialysis. Use of makeup water sources having a low dissolved solids content is another option.
  - Use corrosion inhibitors as a means of sustaining acceptable corrosion rates.
  - Reduce cooling water demand. Use air-cooled exchangers as an alternative to water-cooled heat exchangers.
  - Careful control and optimization of cooling water systems will maximize the number of cycles in a cooling water system, thereby minimizing blowdown.
  - Convert cooling water treatment programs to non-chromate-based treatment as required by EPA regulations, thus eliminating a key source of a toxic metal pollutant (chromium).
  - Use ozone rather than biocides or chlorine to eliminate microorganisms, thereby eliminating potentially toxic chemicals from the refinery wastewater.
- Minimize desalter solids and oil under carry:
  - Improve emulsion formation by using low-shear mixing devices to mix wash water and crude oil and by using low-pressure water to minimize turbulence. Modifications to a desalter are generally not relatively expensive as long as the desalter vessel itself does not have to be replaced; mud rakes can replace water jets to reduce turbulence when removing settled solids; optimizing use of chemical demulsifiers to minimize oil under carry – the project reviewed both the selection of demulsifiers being employed and the quantities used as a function of each crude oil supply source.

6.4   Strategies for reducing flaring

It is widely acknowledged that flaring (Figure 6.15) and venting of associated gas contributes significantly to greenhouse gas (GHG) emissions, with negative impacts on the environment. Associated gas is a blend of hydrocarbons that is released when crude oil is brought to the surface. Gas flaring and venting occur at gas plants, during drilling and testing of oil and gas wells, and from natural gas pipelines during emergencies. Flaring is also performed at refineries

Figure 6.15 A refinery flare.

as a means of incinerating waste gases and for emergency pressure-relief reasons.

While on the surface minimizing the flaring practices at refineries would seem straightforward, the fact is that flares are intended as important safety systems designed to protect site employees, the public, and the refinery assets. More recent thinking has devised state-of-the-art strategies for cost-effective methods that safely minimize if not eliminate the need for flaring at refineries. Some of these strategies are explored in this section.

Before tackling refinery flaring practice, some comments on associated gas related to crude-oil extraction are warranted. Given the increased natural gas prices since the 1970s, governments and industry have recognized the potential economic benefits of using associated gas. It has long been understood that reducing gas flaring and venting practices has direct economic returns. Unfortunately, only a few oil-producing countries have significantly reduced associated gas flaring and venting volumes. In most areas of the world flaring and venting volumes continue to rise with increased oil production. This means that additional incentives such as regulations are needed in order to reduce flaring and venting practices. Within the regulatory arena, the following are key:

- the role of government in defining flaring and venting policies;
- the institutional characteristics of flaring and venting regulation;
- the adopted operational processes and regulatory procedures.

Other relevant factors that affect flaring and venting volumes include:

- the role of standards;
- the impact of financial incentives;
- the effects of contractual rights;
- the structure of the downstream energy markets.

Regulations need to play an important role in achieving reductions in flaring and venting volumes, especially in developing countries. The World Bank Organization and others have noted that governments need to develop strict policies, and legislation, with regulatory agencies independent from regulated operators, to avoid conflicts of interest. Rigorous and efficient operational processes should be adopted, and adequate finance made available to be able to enforce compliance with regulations, while transparent gas flaring and venting application and approval procedures need to be established.

Let us now turn attention to flaring practices at refineries. Flares are combustion devices designed to safely and generally thought to efficiently destroy waste gases generated during the refining process. The practice has been employed since the early days of refining more than a century ago. In refinery operations, flammable waste gases are vented from processing units during normal operation and process upset conditions. These waste gases are collected in piping headers and delivered to a flare system for safe disposal.

A typical flare system often has multiple flares to treat the various sources for waste gases. There may be several different flare types used in a system, depending on site requirements. Flares are primarily safety devices that prevent the release of unburned gas to the atmosphere. Waste gases could burn or even explode if they reached an ignition source outside the refinery, and hence flares are critical from a safe operations standpoint.

When we examine the operational practices of flaring, we come to appreciate that there are two levels of flaring. The first is flaring that occurs during a refinery emergency. This episode may involve a large flow of gases that must be destroyed. Safety is the primary consideration for flaring. During such episodes the flows can be more than a million pounds per hour, depending on the application. The maximum waste-gas flow that can be treated by a flare is referred to as its hydraulic capacity.

The second level of flaring is the treatment of waste gases generated during normal operation, including planned decommissioning of equipment. While safety is still a major focus, control of emissions is the primary reason for this level of practice. The waste-gas flow rate and composition may vary significantly during normal operation. Peterson et al. (2007) have reported typical flare gas compositions, which are summarized in Table 6.1. The values reported show very broad compositions for a wide range of chemical compounds that flares are required to incinerate.

Both the flow rate and composition of the waste gases going to any flare are highly variable. The unsteady flow and variable composition make it difficult to use the waste gases elsewhere in refinery where steady energy demand is required. The variable composition makes it difficult to sell this gas, unless

Table 6.1 Typical refinery flare gas composition

Chemical		Gas composition range (%)			Flare gas (% average)
		Minimum	Maximum	Range	
Methane	CH_4	7.170	82.000	74.830	43.600
Ethane	C_2H_6	0.550	13.100	12.550	3.6600
Propane	C_3H_8	2.040	64.200	62.160	20.300
n-Butane	C_4H_{10}	0.199	28.300	28.101	2.7800
Isobutane	C_4H_{10}	1.330	57.600	56.270	14.300
n-Pentane	C_5H_{12}	0.008	3.3900	3.3820	0.2660
Isopentane	C_5H_{12}	0.096	4.7100	4.6140	0.5300
neo-Pentane	C_5H_{12}	–	0.3420	0.3420	0.0170
n-Hexane	C_6H_{14}	0.026	3.5300	3.5040	0.6360
Ethylene	C_2H_4	0.081	3.2000	3.1190	1.0500
Propylene	C_3H_6	–	42.500	42.500	2.7300
1-Butene	C_4H_8	–	14.700	14.700	0.6960
Carbon monoxide	CO	–	0.9320	0.9320	0.1860
Carbon dioxide	CO_2	0.023	2.8500	2.8270	0.7130
Hydrogen sulfide	H_2S	–	3.8000	3.8000	0.2560
Hydrogen	H_2	–	37.600	37.600	5.5400
Oxygen	O_2	0.019	5.4300	5.4110	0.3570
Nitrogen	N_2	0.073	32.200	32.127	1.3000
Water	H_2O	–	14.700	14.700	1.1400

Source: Peterson et al. (2007).

a costly purification system is added to produce a more consistent composition within the specifications that customers will accept. A major concern is that the waste gases tend to have a low heating value, which means that equipment such as burners must be properly designed for the low calorific value.

Often the waste gases are off-spec product that is being flared because it cannot be sold and is not easily reprocessed to produce on-spec product. Off-spec flaring may occur during transient startup periods until the product is within specification. Also, waste gas pressure is generally low; thus, a compressor is needed to assist in transporting the gases. In most refineries the fuel gas is at a high enough pressure that it can be used to entrain the air needed for combustion so that the burners do not require a fan or blower. Additional piping may be needed to connect the waste gas to the fuel-gas system.

Because there are different types of flares and flare systems, not all flares have the same level of destruction efficiency, and hence although typical reported combustion efficiencies can be 95% or greater, this depends to a large extent on the burner tip design, operating conditions, and the prevailing weather conditions. There are several performance parameters that determine how efficient flaring operations are. One of these is the *smokeless capacity.*

Smokeless capacity is defined as the maximum flow of waste gases that can be sent to the flare without producing significant levels of smoke. Smoke is a visual confirmation of products of incomplete combustion. A flare is typically sized so that the smokeless capacity is at least as much as the maximum waste-gas flow rate expected during normal operation.

Another performance parameter that impacts on combustion efficiency is the *thermal radiation* generated by the flare. The more intense the flame, the higher the combustion efficiency, and hence the greater the thermal radiation. For good combustion efficiency, a flare must have strong thermal radiation. The thermal radiation is a function of the waste-gas flow rate and composition, and is impacted by the flare tip nozzle design. The radiation levels at ground level must be limited in order to avoid injuring personnel and damaging equipment in close proximity to the flare. Flares are best positioned at remote locations within the refinery property line. The height of the flare stack is established such that the acceptable radiation levels are not exceeded at ground level.

In addition to thermal destruction efficiency, consideration is needed of *noise.* Excessive noise can injure personnel, equipment, and property both inside and outside the refinery. Hence the siting issues surrounding flares must take into consideration noise levels and distances to sensitive receptors.

Flares are a source of emissions that include nitrogen oxides (NO_x), sulfur oxides (SO_x), greenhouse gases (CO_2 and CO), and volatile organic compounds (VOCs). These emissions, in combination with any unburned hydrocarbons, contribute to the total refinery emissions.

Flare emissions are generally assumed to be low. AP-42 cites low emission factors, but these factors are based on industry self-reporting, which we believe not to be entirely reliable, largely because in general, emissions from flares have historically been difficult to measure. Conceptually, nearly all flares burn in the open. This means that there is no combustion chamber with a well-contained exhaust stream to insert probes into for extractive or in situ emissions measurements as can be done in an incinerator or a boiler. It is only rather recently that research attention (URS Corp., 2004) has focused on remote monitoring analyzers to measure flare emissions.

There are several technical challenges to monitoring emissions from flares. Both the size of flare flames and elevation above the ground make it difficult to use a hood to collect exhaust gases and measure emissions with a sampling instrument. Another problem is that weather conditions, the waste-gas flow rate, and composition are generally so variable that it is a major challenge to obtain any concentration measurements under steady-state operating conditions. Meteorological conditions also have a direct impact on the performance of a flare. Further, waste-gas flows at high rates, such as those that could occur during emergency pressure let-down conditions, are next to impossible to test in a refinery because for the most part they are rare events. While there are flare-testing facilities capable of simulating very high flow rates, these are not capable of testing for the maximum flow rate that could occur at a refinery.

The above considerations leave us with little confidence in emission factors reported for flaring practices. For those readers that are still not convinced, we note further that the estimated emissions from flares are generally based on measurements obtained under near quiescent conditions, i.e. with little or no wind and never any precipitation. Emissions may be much higher under windy conditions.

In addition we note that many flares use steam as an assist medium to increase air entrainment into the flame to increase the smokeless capacity. Over-steaming (i.e. supplying excessive steam to a flare compared to the waste-gas flow rate), reduces the thermal destruction efficiency. The cooling effect from the use of excessive steam inhibits dispersion of the flared gases, particularly during weather inversions. In the extreme case, over-steaming can actually snuff out the flame and allow waste gases to go into the atmosphere unburned (Banerjee et al., 1985). Most steam-assisted flares are operated in a manual mode. In other words, the steam flow rate is manually controlled and sometimes set for the maximum expected waste-gas flow during normal operation. This unto itself means the flare could be severely over-steamed during periods where the waste-gas flow is much lower.

There is both awareness and momentum in reducing emissions from flaring practices. The Bay Area Air Quality Management District in California established Regulation 12, Rule 12, entitled "Flares at Petroleum Refineries" on 20 July 2005. This rule requires flare minimization projects and requires studies to be performed for area refineries. There is concern that emissions of VOCs from flares may be much higher than previously reported and a recognition that calculation methods defined in AP-42 are not precise (see, for example, the publication by Levy et al., 2006).

Finally, although we have noted the variable and poor quality of waste-gas streams from refineries, the fact is that refineries and chemical plants that rely on flaring practices beyond emergency situations are simply wasting energy and generating pollution. Therefore, industry should be challenged to find ways to recover the gases, either for use in the plant or to sell the refined waste into energy markets.

There are several strategies for minimizing flaring. One strategy is through modifications to plant operational practices. This involves controlling the processes producing waste gases using existing refinery hardware. A low-cost practice is to implement an LDAR program to ensure that equipment is properly maintained to minimize leaks into the waste-gas header. A second strategy is to focus personnel attention towards gaining an understanding of the conditions to be avoided that result in waste gases.

A higher-cost strategy is to invest in new equipment. By investing into hardware that reduces the amount of waste gases going to the flare, the refinery will operate at a greater level of efficiency. An example is to focus on redesign aspects that minimize waste-gas production such as recycling waste gases back into the process or investing in alternative technologies that produce less waste. Still another example is the use of flare gas recovery units (FGRUs). FGRUs can capture waste gases that would have been flared, either for use in the refinery or for cleanup and polishing, such that they may be sold (see, for example, the article by Fisher and Brennan, 2005).

Fisher and Brennan describe the FGRU as a system that can be installed upstream of the flare to capture some or all of the waste gases before they are flared. They note that the flare gas may in some instances have a substantial heating value and could be used as a fuel within the refinery to reduce the amount of purchased fuel. In certain applications, it may be possible to use the recovered flare gas as feedstock or product instead of purchased fuel. Also, an FGRU reduces the continuous flare operation, which subsequently reduces the associated smoke, thermal radiation, noise, and pollutant emissions associated with flaring. Capturing waste gases may also reduce odor levels. Reduced flaring also reduces steam consumption for steam-assisted flares and can extend the service life of the flare tips. In refineries with excess process-generated waste gas beyond fuel gas requirements, an FGRU can also provide a means to scrub the hydrogen sulfide (H_2S) before the clean gas is flared.

When the recovered flare gas is to be utilized as a fuel and the flow is less than or equal to the capacity of the FGRU, the flare gas can be recovered and directed to the refinery fuel-gas header. During these periods, there will be little or no visible flame at the flare. When the flare-gas flow rate exceeds the capacity of the FGRU, the excess flare gas will flow through a liquid seal drum to the flare tip, where it will be combusted. From flaring rates just above the FGRU capacity to a maximum flaring episode, a liquid seal drum can be used in steady-state operation of the flare tip.

FGRU systems can be operated at a slight positive pressure to prevent air infiltration into the system that could create a flammable mixture. The basic processes used in the FGRU are compression and physical separation. The operation is made up of the following steps:

- Process vent gases are recovered from the flare header.
- Gas compressors boost the pressure of the gas.
- Recovered gas is discharged to a service liquid separator.
- The separated gas is passed through a condenser, where the easily condensed constituents may be returned as liquid feedstock while the components that do not easily condense can be returned for use as fuel gas after scrubbing for H_2S and other contaminants.

Gas compression can be performed by compressors depending on the application. For example, if a liquid-ring compressor is used, then separating recovered vapor phase from a mixed liquid is accomplished using a horizontal separator vessel. As the flare gas flows into the header, an established hydrostatic head in the liquid seal drum prevents the flare gas from flowing to the flare. This will result in a slight increase of pressure in the flare gas header, but not enough to significantly affect the capacity of the overpressure protection devices in the refinery. When the flare-gas header pressure reaches the gas recovery initialization setpoint in a batch operation mode, the compression system will begin to compress the flare gas. The FGRU will start and stop with control signals from the pressure-relief controller (PLC).

In contrast, in continuous-operational mode with varying flare loads, parallel compressors can be automatically staged on or off to augment the capacity of a base-load compressor as needed. Based on the inlet pressure of the flare gas header, fine-tuning of FGRU capacity control can be accomplished by recycling recovered gas from the service liquid separator back to the suction side. Discharge of the liquid-ring compressors will flow into the service liquid separator vessel, where the gas and service liquid are disengaged and the compressed recovered flare gas is delivered to the facility fuel gas scrubbing and distribution system. The compressor service liquid, usually water, is used in the compressor as a seal between the rotor and the compressor case. The service liquid is separated from the recovered gas stream, cooled, and recirculated to the gas compressor train for reuse.

Peterson et al. (2007) provide a simplified process diagram for the system described above. They note that the gas-processing capacity of the FGRU adjusts to maintain a positive pressure on the flare header upstream from the existing liquid seal drum. This positive pressure is a safeguard that ensures that air will not be drawn into either the flare system or the FGRU. They further note that if the volume of flare gas that is relieved into the flare system exceeds the capacity of the FGRU, the pressure in the flare header will increase until it exceeds the backpressure exerted on the header by the liquid seal. In this event, excess gas volume will pass through the liquid seal drum and on to the flare, where it will be burned. This mode occurs when there is a rapid increase in flare-gas flow due to an emergency release. Since the liquid seal serves as a backpressure control device for the FGRU, a properly designed deep-liquid seal is critical to the stable operation of the FGRU and flare. As the flow transitions to the flare, this must be done with a very stable liquid level or else unstable flare header pressure could result, affecting FGRU control and proper flare operation.

Gough (2004) discusses the economics and operational experiences with commercialized FGRUs. Installed systems are automated such that there are minimal requirements for direct operator intervention. Gough notes that there are favorable economics, where the payback on the equipment was short enough to justify the capital cost. Such systems were sized to collect most, but not all, of the waste gases. He further cites examples where flaring practices have been reduced by more than 95% since 1997.

6.5 Sulfur recovery strategies

Sulfur recovery refers to the conversion of hydrogen sulfide (H_2S) to elemental sulfur. We may view this process both as a pollution control and a technology that recovers a valuable by-product from a polluting stream. Hydrogen sulfide is a by-product of processing natural gas and refining high-sulfur crude oils. As noted earlier the most common conversion method used is the Claus process. Approximately 90–95% of recovered sulfur is produced by the Claus process. The Claus process typically recovers 95–97% of the hydrogen sulfide feedstream.

The Claus process consists of multistage catalytic oxidation of hydrogen sulfide. Each catalytic stage consists of a gas reheater, a catalyst chamber, and a condenser. The process involves burning one-third of the H_2S with air in a reactor furnace to form sulfur dioxide (SO_2). The furnace normally operates at combustion chamber temperatures ranging from 980 to 1540°C (1800–2800°F) with pressures in the region of 70 kilopascals (kPa; 10 pounds per square inch absolute).

Prior to entering a sulfur condenser, hot gas from the combustion chamber is quenched in a waste heat boiler that generates high- to medium-pressure steam. About 80% of the heat released could be recovered as useful energy, which is a good pollution prevention practice. Liquid sulfur from the condenser runs through a seal leg into a covered pit, from which it is pumped to trucks or railcars for shipment to customers. Approximately 65–70% of the sulfur is recovered at this stage. The cooled gases exiting the condenser are then sent to the catalyst beds. It is here that the remaining uncombusted two-thirds of the hydrogen sulfide undergoes Claus reaction (reacts with SO_2) to form elemental sulfur.

The catalytic reactors operate at lower temperatures, ranging from 200 to 315°C (400–600°F).

Alumina or bauxite is sometimes used as a catalyst. Because this reaction represents an equilibrium chemical reaction, it is not possible for a Claus plant to convert all the incoming sulfur compounds to elemental sulfur. Consequently, at least two and sometimes more stages are used in series to recover the sulfur. Each catalytic stage can recover half to two-thirds of the incoming sulfur. The number of catalytic stages depends upon the level of conversion desired. About 95–97% overall recovery can be achieved depending on the number of catalytic reaction stages and the type of reheating method used.

When a sulfur recovery unit is located in a natural gas processing plant, the type of reheat employed is either auxiliary burners or heat exchangers, with steam reheat being used occasionally. If the sulfur recovery unit is located in a crude oil refinery, the typical reheat scheme uses 3536–4223 kPa (500–600 pounds per square inch guage, psig) steam for reheating purposes. Most plants typically operate with two catalytic stages, although some air quality jurisdictions require three. From the condenser of the final catalytic stage, the process stream passes to some form of tail-gas treatment process.

The tail gas, containing H_2S, SO_2, sulfur vapor, and traces of other sulfur compounds formed in the combustion section, escapes with the inert gases from the tail end of the plant. It is often necessary to follow the Claus unit with a tail-gas cleanup unit to achieve higher recovery. In addition to the oxidation of H_2S to SO_2 and the reaction of SO_2 with H_2S in the reaction furnace, many other side reactions can and do occur in the furnace.

The US EPA's AP-42 reports emission factors and recovery efficiencies for modified Claus sulfur recovery plants. Emissions from the Claus process are directly related to the recovery efficiency.

The efficiency depends upon several factors, including the number of catalytic stages, the concentrations of H_2S and contaminants in the feedstream, stoichiometric balance of gaseous components of the inlet, operating temperature, and catalyst maintenance. Older plants or very small Claus plants have varying sulfur recovery efficiencies.

According to AP-42, at normal operating temperatures and pressures, the Claus reaction is thermodynamically limited to 97–98% recovery. Tail gas from the Claus plant still contains 0.8–1.5% sulfur compounds.

Existing new source performance standards limit sulfur emissions from Claus sulfur recovery plants of greater than 20.32 Mg (22.40 tons) per day capacity to 0.025% by volume (250 parts per million volume, ppmv). This limitation is effective at 0% oxygen on a dry basis if emissions are controlled by an oxidation control system or a reduction control system followed by incineration. This is comparable to the 99.8–99.9% control level for reduced sulfur.

Emissions from the Claus process may be reduced by: (1) extending the Claus reaction into a lower-temperature liquid phase; (2) adding a scrubbing process to the Claus exhaust stream; or (3) incinerating the hydrogen sulfide gases to form sulfur dioxide.

There are five processes available that extend the Claus reaction into a lower-temperature liquid phase, including the BSR/selectox, Sulfreen, Cold Bed Absorption, Maxisulf, and IFP-1 processes. These processes take advantage of the enhanced Claus conversion at cooler temperatures in the catalytic stages. All of these processes give higher overall sulfur recoveries of 98–99% when following downstream of a typical two- or three-stage Claus sulfur recovery unit, and therefore reduce sulfur emissions.

Another approach to reducing sulfur emissions is by adding a scrubber at the tail end of the plant. There are essentially two types of tail-gas scrubbing processes: oxidation tail-gas scrubbers and reduction tail-gas scrubbers. The first scrubbing process is used to scrub SO_2 from incinerated tail gas and recycle the concentrated SO_2 stream back to the Claus process for conversion to elemental sulfur. There are at least three oxidation scrubbing processes: the Wellman–Lord, Stauffer Aquaclaus, and IFP-2. Only the Wellman–Lord process has been applied successfully to US refineries. The Wellman–Lord process uses a wet generative process to reduce stack gas sulfur dioxide concentration to less than 250 ppmv and can achieve approximately 99.9% sulfur recovery. Claus plant tail gas is incinerated and all sulfur species are oxidized to form SO_2 in the Wellman–Lord process. Gases are then cooled and quenched to remove excess water and to reduce gas temperature to absorber conditions. The rich SO_2 gas is then reacted with a solution of sodium sulfite (Na_2SO_3) and sodium bisulfite ($NaHSO_3$) to form the bisulfate. The off-gas is reheated and vented to the atmosphere. The resulting bisulfite solution is boiled in an evaporator–crystallizer, where it decomposes to SO_2 and water (H_2O) vapor and sodium sulfite is precipitated.

The sulfite crystals are separated and redissolved for reuse as lean solution in the absorber. The wet SO_2 gas is directed to a partial condenser, where most of the water is condensed and reused to dissolve sulfite crystals. The enriched SO_2 stream is then recycled back to the Claus plant for conversion to elemental sulfur.

In the second type of scrubbing process, sulfur in the tail gas is converted to H_2S by hydrogenation in a reduction step. After hydrogenation, the tail gas is cooled and water is removed. The cooled tail gas is then sent to the scrubber for H_2S removal prior to venting. There are several reduction scrubbing processes developed for tail-gas sulfur removal: Beavon, Beavon MDEA, SCOT, and ARCO. In the Beavon process, H_2S is converted to sulfur outside the Claus unit using a lean H_2S-to-sulfur process (the Strefford process). The other processes utilize conventional amine scrubbing and regeneration to remove H_2S and recycle back as Claus feed.

Emissions from the Claus process may also be reduced by incinerating sulfur-containing tail gases to form sulfur dioxide. In order to properly remove the sulfur, incinerators must operate at a temperature of 650°C (1200°F) or higher if all the H_2S is to be combusted. Proper air-to-fuel ratios are required to eliminate pluming from the incinerator stack. Stacks should be equipped with analyzers to monitor the SO_2 level.

6.6 Strategies for emissions testing programs

The petroleum refining industry is subject to the different air quality standards listed in Table 6.2.

Table 6.2 Summary of key regulations refineries are subject to

Standard	Sources covered by standard	Parameters or pollutants requiring control
40 CFR 60 Subpart J: Standards of Performance for Petroleum Refineries	Fuel-gas system, incinerators, combustion sources, sulfur recovery units, fluidized catalytic cracking units, and FCCU catalyst regenerator	Volumetric flow, particulate matter (PM), sulfur oxides (SO_x), carbon monoxide (CO), hydrogen sulfide (H_2S), total reduced sulfur (TRS), and opacity
40 CFR 60 Subparts K, KA, and KB: Standards of Performance for Volatile Organic Liquid Storage Vessels	Storage vessels	Volatile organic compounds (VOCs)
40 CFR 60 Subpart GGG: Standards of Performance for Equipment Leaks of VOC in Petroleum Refineries	Compressors, valve, pump, pressure-relief device, sampling connection system, open-ended valve or line, and flange or other connector in VOC service	Total hydrocarbons (THCs)
40 CFR 60 Subpart NNN: Standards of Performance for VOC Emissions from SOCMI Distillation Operations	Distillation operations	Total organic compounds (TOCs)
40 CFR 60 Subpart GG: Standards of Performance for Stationary Gas Turbines	Stationary gas turbines	Nitrogen oxides (NO_x), SO_2
40 CFR 60 Subpart GGG: Standards of Performance for Equipment Leaks of VOC in Petroleum Refineries	Compressors, valve, pump, pressure-relief device, sampling connection system, open-ended valve or line, and flange or other connector in VOC service	Total hydrocarbons (THCs)
40 CFR 60 Subpart NNN: Standards of Performance for VOC Emissions from SOCMI Distillation Operations	Distillation operations	Total organic compounds (TOCs)
40 CFR 60 QQQ: Standards of Performance for VOC Emissions from Petroleum Wastewater Systems	Drain systems, oil–water separators	VOCs

Continued

Table 6.2 Summary of key regulations refineries are subject to—cont'd

Standard	Sources covered by standard	Parameters or pollutants requiring control
40 CFR 61 Subpart E: National Emission Standards for Mercury	Wastewater treatment plant sludge incinerator	Mercury (Hg)
40 CFR 61 Subpart J: National Emission Standards for Equipment Leaks of Benzene	Compressors, valve, pump, pressure-relief device, sampling connection system, open-ended valve or line, and flange or other connector in benzene service	Benzene as THC
40 CFR 61Subpart V: National Emission Standards for Equipment Leaks (Fugitive Emission Sources)	Compressors, valve, pump, pressure-relief device, sampling connection system, open-ended valve or line, and flange or other connector in benzene service	Benzene as THC
40 CFR 61 Subpart Y: National Emission Standards for Benzene Storage Tanks	Storage vessels	Benzene
40 CFR 61Subpart BB: National Emission Standards for Benzene Emissions from Benzene Transfer Operations	Benzene storage tanks and transfer piping systems	Benzene as THC
40 CFR 61 Subpart FF: National Emission Standards for Benzene Waste Operations	Piping and equipment systems that process benzene waste streams	Benzene

While we are critical of the methodology and approaches relied upon in accounting for emissions from the standpoint that calculations are more prevalent than actual monitoring, it is important to recognize that the standards or initiatives listed in Table 6.2 cover virtually every point or fugitive emission source at a refinery and do require approximately 40 different emission testing methods for compliance demonstration. But as we have argued with examples, the frequency of testing, assumptions often applied, and practices in accounting and documentation of emissions do introduce significant errors and an under-reporting of emissions.

Emission testing is among the most challenging of environmental measurement disciplines. It is a critical component of a refinery's compliance strategy.

As such, it is crucial that emission test programs be properly planned and performed if a facility is to generate accurate data that not just simply are defensible and demonstrate compliance with a complex matrix of regulatory requirements, but enable effective strategies to be devised to mitigate potential harm to the public and the environment.

The proper planning and execution of an emission monitoring program requires an understanding of program objectives and the right combination of test methodology, expertise, process operations, and regulatory agency coordination. The following provides some general guidance to establishing strategies for implementing effective emission testing programs.

At the start, the facility should define and establish the purpose of its testing program. It is common sense that if the objectives and reasons for performing certain testing are well defined at the start, then protocols, test methods, procedures, anticipated sampling, and measurement problems while in the field can be understood and considered through proper planning. These considerations will reduce the time and expense associated with test programs, and further help to better define the accuracy of test measurements. It is important to recognize that refinery emission testing programs can be conducted for a number of reasons, including but certainly not limited to:

- performing engineering evaluations, including collection of air pollution control device (APCD) design or vendor guarantee data;
- meeting New Source Performance Standards (NSPS), Maximum Achievable Control Technology (MACT), or Consent Decree (CD) Compliance Demonstration.
- performance specification testing (PST) of continuous emission monitoring systems (CEMS).

It is the intended use of the data that drives the structure of the emission testing program and in turn the structure may vary with each test program focus. For example, a PST or APCD demonstration program might be performed to establish compliance with an NSPS, MACT, or CD compliance demonstration requirement along with specific aspects of the vendor's guarantee. Similarly, evaluation of an APCD in terms of NSPS or MACT may focus on achievement of the emission standard at, say, 95% of rated capacity; however, the vendor's guarantee may include additional operational criteria that need to be considered. These criteria might be pollutant loading or different operating capacities as examples. Such test programs must be designed to include the range of process conditions, test methodology, number of measurements or tests to ensure statistical significance, data evaluation criteria, and reporting requirements that satisfy all objectives. There are qualified emission testing firms that can assist refinery environmental managers with the design of emission test programs that focus on multiple objectives. We do recommend that refinery environmental managers contact emission testing firms during the planning phase to better define and develop a project approach that meets the range of test program objectives.

Ensuring that testing is performed under representative operating conditions can pose unique challenges in establishing a test program. Furthermore, establishing and maintaining proper test conditions to ensure reproducibility poses additional challenges that can impact on cost. In some cases, there may be only a few opportunities throughout a year to conduct testing for a specialized production scenario or special product. Feedstocks may require stockpiling, additional energy may be required for the process, APCD may have to be serviced or operated under different conditions for the test, or it may take several hours to reach steady-state operating conditions in order to obtain representative emissions factors. Market demand for a product may be low at the time testing is required, resulting in additional product storage costs. It is important to match the test conditions to the test program's purpose. Although emission testing programs can be costly, testing program costs generally tend to be low in comparison to the cost of operating the refinery at test conditions. To minimize costs facilities should rely on an emission testing firm that recognizes that the refinery's investment in the testing program significantly exceeds the contract value of the testing.

As part of devising the appropriate strategy, proper selection of test methods should be a key consideration. There are roughly 40 different EPA emission test methods that apply to refineries, with specific applications defined under NSPS, MACT, or the refinery's operating permits. The relationship between air quality regulations and emission test varies between test methods and regulations for other environmental media. Emission standards have been established using empirical data derived from specific test methods. It is important that compliance demonstration testing be conducted in full accordance with the test method that was used for standards development. It is important to note that a reason why test methods have changed very little since publication in the 1970s is because of the connection between the standard setting process and the test method. It is generally argued that changes to test methods could alter the compliance status of stationary sources and undermine the integrity of the regulatory system. Nonetheless, test methods do change and those changes must be incorporated into test programs.

An important recent change in test methods involved revision of EPA instrumental test methods for the measurement of oxygen, carbon dioxide, sulfur dioxide, nitrogen oxides, and carbon monoxide. Although the core measurement methodology for these methods has not been altered, there are significant changes that affect sample location and measurement range. These revised methods became effective on 14 August 2006 and must be used after that date unless there is a specific exemption from the agency with source oversight. Failing to use the current revision of these test methods for test programs after 14 August 2006 is tantamount to using the wrong test method. Maintaining expertise in the specific application of emission testing methods is typically beyond the responsibility of most refinery environmental managers. However, proper application and execution of emission testing methods is central to a refinery's compliance strategy. For this reason, refinery environmental managers are best served by working with emission testing firms that keep them apprised of, or can knowledgably discuss, changes or developments in emission testing methodology.

Another important factor to consider in establishing a strategy is proper development of the test schedule. Once the test program purpose, conditions, and test methods have been established, attention should be given to the test program schedule. This can be an iterative process requiring the refinery environmental manager to balance production schedules, vacation schedules, agency notification requirements, regulatory deadlines, and test team availability. Responsive emission testing firms will be able to assist with the documentation of schedules and the development of multiple scheduling scenarios or options that support test program objectives.

Communication is also a key element in devising the test program strategy. The test program purpose, test conditions, test methods, and schedule should be prepared as a written plan. Elements of the plan should include:

- a program matrix showing sources, test conditions, parameters, and test methods;
- a schedule comprising a Gantt chart showing overall schedule and a table for daily schedules of activities;
- an assignment matrix, which may be at a minimum a table showing program personnel, contact information, program assignment, and project tasks;
- a process operations summary, which may be organized as a table presenting test conditions including target load, time to steady state, test duration, anticipated source conditions (temperature, moisture, flow, analyte concentrations), and documentation requirements;
- definition of data processing and reporting. In this part of the plan the preparer should describe the QC requirements, calculation procedures, and data presentation procedures for each program component (i.e. source and analyte);
- a summary of guidelines and supporting data, which may include excerpts from permits or previous test reports.

It is important that the documented plan be communicated to all responsible parties and team members participating in testing. Upfront effort in developing relevant test plan information for program participants will pay off when comes the time for test program execution because program participants will have a clear understanding of their responsibilities, how their responsibilities affect or relate to the work of others, and defining the role they play in completing a successful emission test program.

Site preparation considerations are another important factor. In the past, the primary concern about sample location was related to volumetric flow measurement, with secondary consideration being given to source-gas stratification issues. Revised instrumental test methods provide specific, quantitative requirements for stratification. As a result, the sampling strategy – and the time required for a test – changes with the degree of stratification at the sample location. It is therefore important to ensure that sample locations are selected that meet the requirements of the method and allow for efficient testing. An additional factor to consider relative to site preparations is specific test contractor requirements. Testing firms rely on different brands of test equipment and use custom-built mobile laboratories. As such, different test companies may require different port sizes, different monorail supports, or different power requirements. Failure to recognize these differences can result in last test program delays. Site preparation requirements are determined by both test method and testing firm. Advance inspections will allow the refinery and the test team to properly prepare for the upcoming test or rearrange the overall test program to accommodate extensive efforts associated with the preparation of problem test locations.

Once the on-site work begins, it is important to review progress on a daily basis to ensure that the test program stays on schedule, that the proper sources were tested under the proper test conditions, and that the proper supporting data were collected by operations personnel. Many of the test methods used at refineries provide results in the field. It is important to review flow values, concentration data and, in the case of PST programs, the relative accuracy calculations, each test day. This daily review of progress, data, and documentation will help ensure proper completion of the test program and timely reporting.

A final consideration is any preplanning that will minimize or eliminate delays on reporting results. The most effective way to eliminate test report delays is to insist on the delivery of a draft report in template form before the test team arrives to conduct the test. The template should include all elements, including data table numbers and titles. This is not simply an outline, but rather a skeletal report. For example, data tables should include source identification data and indicate permit limits and reporting units. This document will assist management in understanding the planned organization and presentation of data and facilitate revisions early in the process. The template will also help the test team secure all required process or refinery information and will focus a significant part of the reporting effort on the front end of the project.

Emissions test programs are entirely designed to meet compliance obligations. The regulations are after all the drivers for reducing emissions. However, testing can also afford the refinery an opportunity to identify pollution prevention opportunities, even though these may not necessarily be among the objectives of the program. Emissions test programs should not deviate from their intended purpose, but they can afford an opportunity to screen data that could be useful for devising separate activities under different strategies to assess pollution prevention opportunities. Therefore, as a final recommendation we advise that the strategy take into consideration that possible areas of overlapping test data might serve to define further test programs aimed at identifying source reduction opportunities.

References

Banerjee, K., Cheremisinoff, N.P., Cheremisinoff, P.N., 1985. Flare Gas Systems Pocket Handbook. Gulf Publishing, Houston, TX.

Design Guidelines for Flare Gas Practices. In: Cheremisinoff, N. (Ed.), Handbook of Heat and Mass Transfer, Vol. 1. Gulf Publishing, Houston, TX, pp. 1401–1433. Chap. 43.

Fisher, P.W., Brennan, D., 2005. Minimize Flaring with Flare Gas Recovery. Hydrocarbon Processing 81 (May), 83–85.

Gough, R., 2004. Flint Hills Resources Shows Flare for Not Flaring. World Refining 14 (6), 36–39.

Levy, R., Randel, L., Healy, M., and Weaver, D., 2006. Reducing Emissions from Plant Flares. Proceedings of the Air and Waste Management Association Conference and Exhibition, New Orleans, LA, June, Paper #618.

Peterson, J., Tuttle, N., Cooper, H., Baukal, C., 2007. Minimize Facility Flaring. www. HydrocarbonProcessing.com. Hydrocarbon Processing, 111–115.

Corp, U.R.S., 2004. Passive FTIR Phase I Testing of Simulated and Controlled Flare Systems – Final Report. Prepared for the Texas Commission on Environmental Quality June.

US Environmental Protection Agency (EPA), 1995. Profile of the Petroleum Refining Industry, EPA 310-R-95-013. US Government Printing Office, Washington, DC. September.

Appendix

The black rectangle represents a covered sump in the Los Cabos Santa Maria neighborhood. Illustration prepared by URS.

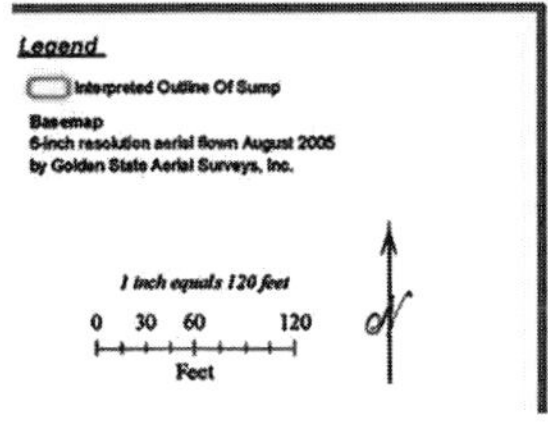

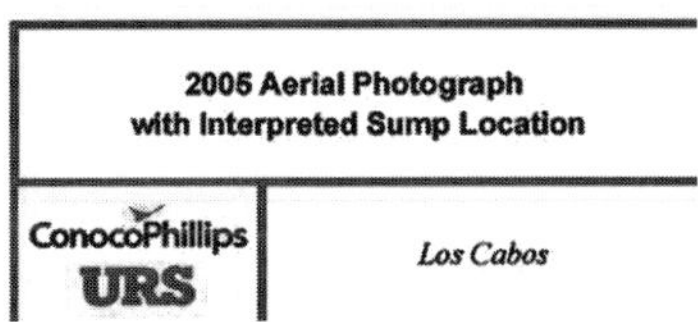

OSHA Permissible Exposure Limits (PEL)

CAS No. (c)	CHEMICAL NAME	ppm (a)(1)	mg/m^3 (b)(1)	Skin designation
67-64-1	Acetone	1000	2400	
7664-41-7	Ammonia	50	35	
7440-36-0	Antimony and compounds (as Sb)		0.5	
7440-39-3	Barium, soluble compounds (as Ba)		0.5	
106-99-0	Butadiene	1 ppm/5 ppm STEL		
124-38-9	Carbon dioxide	5000	9000	
630-08-0	Carbon monoxide	50	55	
56-23-5	Carbon tetrachloride		(2)	
7782-50-5	Chlorine	(C)1	(C)3	
	Chromic acid and chromates (as CrO(3))		(2)	
7440-47-3	Chromium (II) compounds (as Cr)		0.5	
7440-47-3	Chromium (III) compounds (as Cr)		0.5	
7440-47-3	Chromium metal and insol. salts (as Cr)		1	
7440-48-4	Cobalt metal, dust, and fume (as Co)		0.1	
7440-50-8	Copper fume (as Cu)		0.1	
7440-50-8	Copper dusts and mists (as Cu)		1	
Varies	Cresol (all isomers)	5	22	X
98-82-8	Cumene	50	245	X
110-82-7	Cyclohexane	300	1050	
100-41-4	Ethyl benzene	100	435	
106-93-4	Ethylene dibromide		(2)	
107-06-2	Ethylene dichloride (1,2-Dichloroethane)		(2)	

OSHA Permissible Exposure Limits (PEL)—cont'd

CAS No. (c)	CHEMICAL NAME	ppm (a)(1)	mg/m^3 (b)(1)	Skin designation
7664-39-3	Hydrogen fluoride		(2)	
7783-06-4	Hydrogen sulfide		(2)	
7439-96-5	Manganese compounds (as Mn)		(C)5	
91-20-3	Naphthalene	10	50	
13463-39-3	Nickel carbonyl (as Ni)	0.001	0.07	
7440-02-0	Nickel, metal and insoluble compounds (as Ni)		1	
7440-02-0	Nickel, soluble compounds (as Ni)		1	
108-95-2	Phenol	5	19	X
7664-38-2	Phosphoric acid		1	
7446-09-5	Sulfur dioxide	5	13	
7664-93-9	Sulfuric acid		1	
108-88-3	Toluene		(2)	
1330-20-7	m-Xylenes	100	435	
95-47-6	o-Xylene	100	435	
106-42-3	p-Xylene	100	435	
7646-85-7	Zinc chloride fume		1	
1314-13-2	Zinc oxide fume		5	
1314-13-2	Zinc oxide, total dust		15	
1314-13-2	Zinc oxide, respirable fraction		5	
557-05-1	Zinc stearate, total dust		15	
557-05-1	Zinc stearate, respirable fraction		5	

Reference: 29 CFR Part Number: 1910 Part Title: Occupational Safety and Health Standards Subpart: Z Subpart Title: Toxic and Hazardous Substances, Standard Number: 1910.1000 TABLE Z-1, Title: TABLE Z-1 Limits for Air Contaminants. Available http://www.osha.gov/pls/o

OSHA Permissible Exposure Limits (PEL) Time Weighted Averages

CAS No.	CHEMICAL NAME	8-hour TWA	Acceptable ceiling concentration	Acceptable maximum peak above the acceptable ceiling concentration for an 8-hr shift	Maximum duration
71-43-2	Benzene (a)	10 ppm	25 ppm	50 ppm	10 minutes
56-23-5	Carbon tetrachloride	10 ppm	25 ppm	200 ppm	5 min. in any 3 hrs
1333-82-0	Chromic acid and chromates		1 mg/10 m(3)		
106-93-4	Ethylene dibromide	20 ppm	30 ppm	50 ppm	5 minutes
107-06-2	Ethylene dichloride	50 ppm	100 ppm	200 ppm	5 min. in any 3 hrs
7664-39-3	Hydrogen fluoride	3 ppm			
7783-06-4	Hydrogen sulfide		20 ppm	50 ppm	10 mins once only if no other meas. exp. occurs
100-42-5	Styrene	100 ppm	200 ppm	600 ppm	5 mins in any 3 hrs
127-18-4	Tetrachloroethylene	100 ppm	200 ppm	300 ppm	5 mins in any 3 hrs
108-88-3	Toluene	200 ppm	300 ppm	500 ppm	10 minutes

Footnote (a) This standard applies to the industry segments exempt from the 1 ppm 8-hour TWA and 5 ppm STEL of the benzene standard at 1910.1028.
Reference: 29 CFR Part Number: 1910 Part Title: Occupational Safety and Health Standards Subpart: Z Subpart Title: Toxic and Hazardous Substances Standard Number: 1910.1000 TABLE Z-2 Title: TABLE Z-2. Available http://www.osha.gov/pls/oshaweb/owadisp.show_document?

OSHA Immediately Dangerous to Life and Health (IDLH) Values

CAS No.	CHEMICAL NAME	IDLH (ppm)	STEL/ Ceiling (ppm)
67-64-1	Acetone	2500	1000
7664-41-7	Ammonia	300	35
7440-36-0	Antimony & compounds	50 mg/m^3	
106-99-0	1,3-Butadiene	2000 ++	
124-38-9	Carbon dioxide	40,000	30,000
630-08-0	Carbon monoxide	1200	
7782-50-5	Chlorine	10	1
	Chromium (II) compounds (as Cr)	250 mg/m^3	
	Chromium (III) compounds (as Cr)	25 mg/m^3	
7440-47-3	Chromium metal and insoluble salts	250 mg/m^3	
7440-48-4	Cobalt metal, dust, and fume (as Co)	20 mg/m^3	
1317-38-0	Copper fume (as Cu)	100 mg/m^3	
7440-50-8	Copper dusts and mists (as Cu)	100 mg/m^3	
Varies	Cresol, all isomers	250	
98-82-8	Cumene	900	
110-82-7	Cyclohexane	1300	
100-41-4	Ethyl benzene	800	125
7439-96-5	Manganese compounds (as Mn)	500 mg/m^3	5 mg/m^3+
7439-96-5	Manganese fume (as Mn)	500 mg/m^3	5 mg/m^3+
91-20-3	Naphthalene	250	15
13463-39-3	Nickel carbonyl (as Ni)	2 ++	
7440-02-0	Nickel, metal and insoluble compounds (as Ni)	10 mg/m^3++	
108-95-2	Phenol	250	
7664-38-2	Phosphoric acid	1000 mg/m^3	
7446-09-5	Sulfur dioxide	100	5
7664-93-9	Sulfuric acid	15 mg/m^3	3 mg/m^3+
Varies	Xylenes (o-, m-, p- isomers)	900	150
7646-85-7	Zinc chloride fume	50 mg/m^3	
1314-13-2	Zinc oxide fume	500 mg/m^3	
1314-13-2	Zinc oxide	500 mg/m^3	
1314-13-2	Zinc oxide	500 mg/m^3	

" + " indicates a Ceiling Value.

" ++ " indicates that the chemical is believed, by NIOSH, to be a potential carcinogen.

**Reference: http://www.labsafety.com/refinfo/ezfacts/ezf232.htm

NIOSH Recommended Exposure Limits (REL)

CAS No.	CHEMICAL NAME	NIOSH REL TWA (ppm)	NIOSH REL TWA (mg/m³)	NIOSH Ceiling REL (ppm)	NIOSH Ceiling REL (mg/m³)	NIOSH REL STEL (ppm)	NIOSH REL STEL (mg/m³)
67-64-1	Acetone	250	590				
7664-41-7	Ammonia	25	18			35	27
7440-36-0	Antimony		0.5				
71-43-2	Benzene	0.1				1	
124-38-9	Carbon dioxide	5000	9000			30,000	54,000
630-08-0	Carbon monoxide	35	40	200	229		
56-23-5	Carbon tetrachloride					2 (60 min)	12.6 (60 min)
7782-50-5	Chlorine			0.5 (15 min)	1.45 (15 min)		
1333-82-0	Chromic acid and chromates		0.001				
	Chromium(II) compounds (as Cr)		0.5				
	Chromium(III) compounds (as Cr)		0.5				
7440-47-3	Chromium metal		0.5				
10210-68-1	Cobalt carbonyl (as Co)		0.1				
16842-03-8	Cobalt hydrocarbonyl (as Co)		0.1				
7440-48-4	Cobalt metal, dust, and fume (as Co)		0.05				
7440-50-8	Copper (dusts and mists, as Cu)		1				
108-39-4	m-Cresol	2.3	10				
95-48-7	o-Cresol	2.3	10				
106-44-5	p-Cresol	2.3	10				
98-82-8	Cumene	50 (skin)	245 (skin)				
110-82-7	Cyclohexane	300	1050				
75-71-8	Dichlorodifluoromethane	1000	4950				
111-42-2	Diethanolamine	3	15				
100-41-4	Ethyl benzene	100	435			125	545
106-93-4	Ethylene dibromide	0.045 (15 min)		0.13 (15 min)			
107-06-2	Ethylene dichloride	1	4			2	8

CAS	Substance						
7664-39-3	Hydrogen fluoride	3 (15 min)	2.5 (15 min)	6 (15 min)	5 (15 min)		
7664-93-9	Hydrogen sulfide			10 (10 min)	15 (10 min)		
7439-92-1	Lead		0.050				
7439-96-5	Manganese compounds (as Mn)		1				3
7439-96-5	Manganese fume (as Mn)		1				3
12079-65-1	Manganese cyclopentadienyl tricarbonyl (as Mn)		0.1 (skin)				
91-20-3	Naphthalene	10	50			15	75
13463-39-3	Nickel carbonyl	0.001	0.007				
7440-02-0	Nickel metal and other compounds (as Ni)		0.015				
108-95-2	Phenol	5 (15 min)	19 (15 min)	15.6 (15 min)	60 (15 min)		
7664-38-2	Phosphoric acid		1				1
100-42-5	Styrene	50	215			100	425
7446-09-5	Sulfur dioxide	2	5			5	13
7664-93-9	Sulfuric acid		1				
127-18-4	Tetrachloroethylene	(minimize occupational exposure)					
108-88-3	Toluene	100	375			150	560
95-63-6	1,2,4-Trimethylbenzene	25	125				
108-38-3	m-Xylene	100	435			150	655
95-47-6	o-Xylene	100	435			150	655
106-42-3	p-Xylene	100	435			150	655
7646-85-7	Zinc chloride fume		1				2
1314-13-2	Zinc oxide		5 (dust) 5 (fume)		15 (dust)		10 (fume)
557-05-1	Zinc stearate		10 (total) 5 (resp)				

Reference: http://www.cdc.gov/niosh/npg/npgname-a.html

NIOSH Immediately Dangerous to Life and Health (IDLH) Values

CAS No.	CHEMICAL NAME	Original IDLH	Revised IDLH
67-64-1	Acetone	20,000 ppm	2500 ppm [LEL]
7664-41-7	Ammonia	500 ppm	300 ppm
7440-36-0 (Metal)	Antimony compounds (as Sb)	80 mg Sb/m^3	50 mg Sb/m^3
7440-39-3 (Metal)	Barium (soluble compounds, as Ba)	1100 mg Ba/m^3	50 mg Ba/m^3
71-43-2	Benzene	3000 ppm	500 ppm
106-99-0	1,3-Butadiene	20,000 ppm [LEL]	2000 ppm [LEL]
124-38-9	Carbon dioxide	50,000 ppm	40,000 ppm
630-08-0	Carbon monoxide	1500 ppm	1200 ppm
56-23-5	Carbon tetrachloride	300 ppm	200 ppm
7782-50-5	Chlorine	30 ppm	10 ppm
1333-82-0 (CrO$_3$)	Chromic acid and chromates	30 mg/m^3 (as CrO$_3$)	15 mg Cr(VI)/m^3
Varies	Chromium (II) compounds [as Cr(II)]	N.E.	250 mg Cr(II)/m^3
Varies	Chromium (III) compounds [as Cr(III)]	N.E.	25 mg Cr(III)/m^3
7440-47-3	Chromium metal (as Cr)	N.E.	250 mg Cr/m^3
7440-48-4 (Metal)	Cobalt metal, dust and fume (as Co)	20 mg Co/m^3	20 mg Co/m^3 [Unch]
7440-50-8 (Metal)	Copper (dusts and mists, as Cu)	N.E.	100 mg Cu/m^3
1317-38-0 (CuO)	Copper fume (as Cu)	N.E.	100 mg Cu/m^3
95-48-7 (o-isomer), 108-39-4 (m-isomer), 106-44-5 (p-isomer)	Cresol (o-, m-, p- isomers)	250 ppm	250 ppm [Unch]
98-82-8	Cumene	8000 ppm	900 ppm [LEL]
110-82-7	Cyclohexane	10,000 ppm	1300 ppm [LEL]
75-71-8	Dichlorodifluoromethane	50,000 ppm	15,000 ppm

NIOSH Immediately Dangerous to Life and Health (IDLH) Values—cont'd

CAS No.	CHEMICAL NAME	Original IDLH	Revised IDLH
100-41-4	Ethyl benzene	2000 ppm	800 ppm [LEL]
106-93-4	Ethylene dibromide	400 ppm	100 ppm
107-06-2	Ethylene dichloride	1000 ppm	50 ppm
7664-39-3	Hydrogen fluoride (as F)	30 ppm	30 ppm [Unch]
7783-06-4	Hydrogen sulfide	300 ppm	100 ppm
7439-92-1	Lead compounds (as Pb)	700 mg Pb/m^3	100 mg Pb/m^3
7439-96-5	Manganese compounds (as Mn)	N.E.	500 mg Mn/m^3
91-20-3	Naphthalene	500 ppm	250 ppm
13463-39-3	Nickel carbonyl (as Ni)	7 ppm	2 ppm
7440-02-0	Nickel metal and other compounds (as Ni)	N.E.	10 mg Ni/m^3
108-95-2	Phenol	250 ppm	250 ppm [Unch]
7664-38-2	Phosphoric acid	10,000 mg/m^3	1000 mg/m^3
100-42-5	Styrene	5000 ppm	700 ppm
7446-09-5	Sulfur dioxide	100 ppm	100 ppm [Unch]
7664-93-9	Sulfuric acid	80 mg/m^3	15 mg/m^3
127-18-4	Tetrachloroethylene	500 ppm	150 ppm
108-88-3	Toluene	2000 ppm	500 ppm
95-47-6	Xylene (o-isomer)	1000 ppm	900 ppm
108-38-3	Xylene (m-isomer)	1000 ppm	900 ppm
106-42-3	Xylene (p- isomers)	1000 ppm	900 ppm
7646-85-7	Zinc chloride fume	4800 mg/m^3	50 mg/m^3
1314-13-2	Zinc oxide	2500 mg/m^3	500 mg/m^3

N.E. = no evidence.

Reference: http://www.cdc.gov/niosh/idlh/intridl4.html

ACGIH Threshold Limit Value (TLV)

CAS No.	CHEMICAL NAME	ppm	mg/m^3
67-64-1	Acetone	750	
7664-41-7	Ammonia	25	
7440-36-0	Antimony & compounds		0.5
	Barium, soluble compounds		0.2
106-99-0	1,3-Butadiene	2	
124-38-9	Carbon dioxide	5000	
630-08-0	Carbon monoxide	25	
7782-50-5	Chlorine	0.5	
	Chromium (III) compounds		0.5
1317-38-0	Copper fume		0.2
7440-50-8	Copper dusts and mists		1
Varies	Cresol, all isomers	5	
98-82-8	Cumene	50	
110-82-7	Cyclohexane	300	
100-41-4	Ethyl benzene	100	
91-20-3	Naphthalene	10	
13463-39-3	Nickel carbonyl (as Ni)	0.05	
7440-02-0	Nickel, metal and insoluble compounds (as Ni)		1
7440-02-0	Nickel, soluble compounds (as Ni)		0.1
108-95-2	Phenol	5	
7664-38-2	Phosphoric acid		1
7446-09-5	Sulfur dioxide	2	
7664-93-9	Sulfuric acid		1
Varies	Xylenes (o-, m-, p- isomers)	100	
7646-85-7	Zinc chloride fume		1
1314-13-2	Zinc oxide fume		5
1314-13-2	Zinc oxide		10

**Reference: http://www.labsafety.com/refinfo/ezfacts/ezf232.htm

ACGIH Immediately Dangerous to Life and Health (IDLH)

CAS No.	CHEMICAL NAME	ppm	mg/m^3
67-64-1	Acetone	20,000	
7664-41-7	Ammonia	500	
7440-36-0 (Metal)	Antimony compounds (as Sb)		80 Sb/m^3
7440-39-3 (Metal)	Barium (soluble compounds, as Ba)		1100 mg Ba/m^3
71-43-2	Benzene	3000	
106-99-0	1,3-Butadiene	20,000 (LEL)	
124-38-9	Carbon dioxide	50,000	
630-08-0	Carbon monoxide	1500	
56-23-5	Carbon tetrachloride	300	
7782-50-5	Chlorine	30	
Varies	Chromium (II) compounds [as Cr(II)]	no evidence	
Varies	Chromium (III) compounds [as Cr(III)]	no evidence	
7440-47-3	Chromium metal (as Cr)	no evidence	
7440-48-4 (Metal)	Cobalt metal, dust and fume (as Co)		20 mg Co/m^3
7440-50-8 (Metal)	Copper (dusts and mists, as Cu)	no evidence	
1317-38-0 (CuO)	Copper fume (as Cu)	no evidence	
95-48-7 (o-isomer), 108-39-4 (m-isomer), 106-44-5 (p-isomer)	Cresol (o-, m-, p- isomers)	250	
98-82-8	Cumene	8000	
110-82-7	Cyclohexane	10,000	
75-71-8	Dichlorodifluoromethane	50,000	
100-41-4	Ethyl benzene	2000	
106-93-4	Ethylene dibromide	400	
107-06-2	Ethylene dichloride	1000	
7664-39-3	Hydrogen fluoride (as F)	30	
7783-06-4	Hydrogen sulfide	300	
7439-92-1	Lead compounds (as Pb)		700 mg Pb/m^3

Continued

ACGIH Immediately Dangerous to Life and Health (IDLH)—cont'd

CAS No.	CHEMICAL NAME	ppm	mg/m^3
7439-96-5	Manganese compounds (as Mn)	no evidence	
91-20-3	Naphthalene	500	
13463-39-3	Nickel carbonyl (as Ni)	7	
7440-02-0	Nickel metal and other compounds (as Ni)	no evidence	
108-95-2	Phenol	250	
7664-38-2	Phosphoric acid		10,000
100-42-5	Styrene	5000	
7446-09-5	Sulfur dioxide	100	
7664-93-9	Sulfuric acid		80
127-18-4	Tetrachloroethylene	500	
108-88-3	Toluene	2000	
95-47-6	Xylene (o-isomer)		1000
108-38-3	Xylene (m-isomer)		1000
106-42-3	Xylene (p- isomers)		1000
7646-85-7	Zinc chloride fume		4800
1314-13-2	Zinc oxide	no evidence	

Reference: http://www.cdc.gov/niosh/idlh/intridl4.html

ATSDR Minimal Risk Levels (MRLs) December 2006

CAS No.	CHEMICAL NAME	Route	Duration	MRL	Factors	Endpoint	Draft/Final	Cover Date
000067-64-1	ACETONE	Inh.	Acute	26 ppm	5	Neurol.	Final	May-94
			Int.	13 ppm	100	Neurol.		
			Chr.	13 ppm	100	Neurol.		
		Oral	Int.	2 mg/kg/day	100	Hemato.		
007664-41-7	AMMONIA	Inh.	Acute	1.7 ppm	30	Resp.	Final	4-Oct
			Chr.	0.1 ppm	30	Resp.		
000120-12-7	ANTHRACENE	Oral	Int.	10 mg/kg/day	100	Hepatic	Final	Aug-95
007440-39-3	BARIUM, SOLUBLE SALTS	Oral	Int.	0.7 mg/kg/day	100	Renal	Draft	5-Sep
			Chr.	0.6 mg/kg/day	100	Renal		
000071-43-2	BENZENE	Inh.	Acute	0.009 ppm	300	Immuno.	Draft	5-Sep
			Int.	0.006 ppm	300	Immuno.		
			Chr.	0.003 ppm	10	Immuno.		
000056-23-5	CARBON TETRACHLORIDE	Inh.	Int.	0.03 ppm	30	Hepatic	Final	5-Sep
			Chr.	0.03 ppm	30	Hepatic		
		Oral	Acute	0.02 mg/kg/day	300	Hepatic		
			Int.	0.007 mg/kg/day	100	Hepatic		
007738-94-5	CHROMIUM(VI), AEROSOL MISTS	Inh.	Int.	0.000005 mg/m^3	100	Resp.	Final	Sep-00
018540-29-9	CHROMIUM(VI), PARTICULATES	Inh.	Int.	0.001 mg/m^3	30	Resp.	Final	Sep-00

Continued

ATSDR Minimal Risk Levels (MRLs) December 2006—cont'd

CAS No.	CHEMICAL NAME	Route	Duration	MRL	Factors	Endpoint	Draft/Final	Cover Date
007440-48-4	COBALT	Inh.	Chr.	0.0001 mg/m^3	10	Resp.	Final	4-Oct
		Oral	Int.	0.01 mg/kg/day	100	Hemato.		
		Rad.	Acute	4 mSv	3	Develop.		
			Chr.	1 mSv/yr	3	Other		
007440-50-8	COPPER	Oral	Acute	0.01 mg/kg/day	3	Gastro.	Final	4-Oct
			Int.	0.01 mg/kg/day	3	Gastro.		
001319-77-3	CRESOLS	Oral	Int.	0.1 mg/kg/day	100	Resp.	Draft	6-Sep
000100-41-4	ETHYLBENZENE	Inh.	Int.	1.0 ppm	100	Develop.	Final	Jul-99
000107-21-1	ETHYLENE GLYCOL	Inh.	Acute	0.5 ppm	100	Renal	Final	Sep-97
		Oral	Acute	2.0 mg/kg/day	100	Develop.		
			Chr.	2.0 mg/kg/day	100	Renal		
007664-39-3	HYDROGEN FLUORIDE	Inh.	Acute	0.02 ppm	30	Resp.	Final	3-Sep
007783-06-4	HYDROGEN SULFIDE	Inh.	Acute	0.07 ppm	27	Resp.	Final	6-Jul
			Int.	0.02 ppm	30	Resp.		
007439-96-5	MANGANESE	Inh.	Chr.	0.00004 mg/m^3	500	Neurol.	Final	Sep-00
001634-04-4	METHYL-T- BUTYL ETHER	Inh.	Acute	2 ppm	100	Neurol.	Final	Aug-96
			Int.	0.7 ppm	100	Neurol.		
			Chr.	0.7 ppm	100	Renal		

CAS Number	Chemical	Route	Duration	Value	UF	Endpoint	Status	Date
		Oral	Acute	0.4 mg/kg/day	100	Neurol.		
			Int.	0.3 mg/kg/day	300	Hepatic		
000091-20-3	NAPHTHALENE	Inh.	Chr.	0.0007 ppm	300	Resp.	Final	5-Sep
		Oral	Acute	0.6 mg/kg/day	90	Neurol.		
			Int.	0.6 mg/kg/day	90	Neurol.		
007440-02-0	NICKEL	Inh.	Int.	0.0002 mg/m^3	30	Resp.	Final	5-Sep
			Chr.	0.00009 mg/m^3	30	Resp.		
000108-95-2	PHENOL	Inh.	Acute	0.02 ppm	30	Resp.	Draft	6-Sep
		Oral	Acute	0.6 mg/kg/day	100	Body Wt.		
000100-42-5	STYRENE	Inh.	Chr.	0.06 ppm	100	Neurol.	Final	Sep-92
		Oral	Int.	0.2 mg/kg/day	1000	Hepatic		
007446-09-5	SULFUR DIOXIDE	Inh.	Acute	0.01 ppm	9	Resp.	Final	Dec-98
000127-18-4	TETRACHLOROETHYLENE	Inh.	Acute	0.2 ppm	10	Neurol.	Final	Sep-97
			Chr.	0.04 ppm	100	Neurol.		
		Oral	Acute	0.05 mg/kg/day	100	Develop.		
000108-88-3	TOLUENE	Inh.	Acute	1 ppm	10	Neurol.	Final	Sep-00
			Chr.	0.08 ppm	100	Neurol.		
		Oral	Acute	0.8 mg/kg/day	300	Neurol.		
			Int.	0.02 mg/kg/day	300	Neurol.		

Continued

ATSDR Minimal Risk Levels (MRLs) December 2006—cont'd

CAS No.	CHEMICAL NAME	Route	Duration	MRL	Factors	Endpoint	Draft/Final	Cover Date
001330-20-7	XYLENES, MIXED	Inh.	Acute	2 ppm	30	Neurol.	Draft	5-Sep
			Int.	0.6 ppm	90	Neurol.		
			Chr.	0.05 ppm	300	Neurol.		
		Oral	Acute	1 mg/kg/day	100	Neurol.		
			Int.	1 mg/kg/day	300	Neurol.		
			Chr.	0.6 mg/kg/day	300	Neurol.		
007440-66-6	ZINC	Oral	Int.	0.3 mg/kg/day	3	Hemato.	Final	5-Sep
			Chr.	0.3 mg/kg/day	3	Hemato.		
000071-55-6	1,1,1-TRICHLOROETHANE	Inh.	Acute	2 ppm	100	Neurol.	Final	6-Jul
			Int.	0.7 ppm	100	Neurol.		
		Oral	Int.	20 mg/kg/day	100	Body Wt.		
000107-06-2	1,2-DICHLOROETHANE	Inh.	Chr.	0.6 ppm	90	Hepatic	Final	1-Sep
		Oral	Int.	0.2 mg/kg/day	300	Renal		

Reference: http://www.atsdr.cdc.gov/mrls/index.html

WHO Air Quality Guidelines

CAS No.	CHEMICAL NAME	Time Weighted Average (TWA)	Averaging Time
630-08-0	Carbon monoxide	100 mg/m^3	15 min
		60 mg/m^3	30 min
		30 mg/m^3	1 hr
		10 mg/m^3	8 hrs
107-06-2	1,2-Dichloroethane	0.7 mg/m^3	24 hrs
7783-06-4	Hydrogen sulfide	150 μg/m^3	24 hrs
7439-92-1	Lead	0.5 μg/m^3	annual
7439-96-5	Manganese	0.15 μg/m^3	annual
100-42-5	Styrene	0.26 mg/m^3	1 week
7446-09-5	Sulfur dioxide	500 μg/m^3	10 min
		125 μg/m^3	24 hrs
		50 μg/m^3	annual
127-18-4	Tetrachloroethylene	0.25 mg/m^3	annual
108-88-3	Toluene	0.26 mg/m^3	1 week

Reference: http://www.euro.who.int/document/e71922.pdf

Contaminant		Toxicity and Chemical-specific Information										
		SFO		IUR		RfDo		RfCi				RAGS Part E
Analyte	CAS No.	(mg/kg-day)$^{-1}$	key	(μg/m^3)$^{-1}$	key	(mg/kg-day)	key	(mg/m^3)	key	voc	muta-gen	GIABS
Acetone	67-64-1					9.0E-01	I	3.1E+01	A	V		1
Ammonia	7664-41-7						H	1.0E-01	I			1
Antimony (metallic)	7440-36-0					4.0E-04	I					0.15
Antimony Pentoxide	1314-60-9					5.0E-04	H					0.15
Antimony Potassium Tartrate	304-61-0					9.0E-04	H					0.15
Antimony Tetroxide	1332-81-6					4.0E-04	H					0.15
Antimony Trioxide	1309-64-4					4.0E-04	H	2.0E-04	I			0.15
Barium	7440-39-3					2.0E-01	I	5.0E-04	H			0.07
Benzene	71-43-2	5.5E-02	I	7.8E-06	I	4.0E-03	I	3.0E-02	I	V		1
Biphenyl, 1,1'-	92-52-4					5.0E-02	I			V		1
Butadiene, 1,3-	106-99-0			3.0E-05	I			2.0E-03	I	V		1
Carbon Tetrachloride	56-23-5	1.3E-01	I	1.5E-05	I	7.0E-04	I	1.9E-01	A	V		1
Chlorine	7782-50-5					1.0E-01	I	1.5E-04	A			1
Chromium (III) (insoluble salts)	16065-83-1					1.5E+00	I					0.01
Chromium VI (chromic acid mists)	18540-29-9			1.2E-02	I	3.0E-03	I	8.0E-06	I			1
Chromium VI (particulates)	18540-29-9			1.2E-02	I	3.0E-03	I	1.0E-04	I			0.03
Copper	7440-50-8					4.0E-02	H					1
Cresol, m-	108-39-4					5.0E-02	I					1
Cresol, o-	95-48-7					5.0E-02	I					1
Cresol, p-	106-44-5					5.0E-03	H					1
Cumene	98-82-8					1.0E-01	I	4.0E-01	I	V		1
Cyclohexane	110-82-7							6.0E+00	I	V		1
Dibromoethane, 1,2-	106-93-4	2.0E+00	I	6.0E-04	I	9.0E-03	I	9.0E-03	I	V		1
Dichlorodifluoro-methane	75-71-8					2.0E-01	I	2.0E-01	H	V		1
Dichloroethane, 1,2-	107-06-2	9.1E-02	I	2.6E-05	I			2.4E+00	A	V		1
Ethylbenzene	100-41-4	1.1E-02	C	2.5E-06	C	1.0E-01	I	1.0E+00	I	V		1
Ethylene Glycol	107-21-1					2.0E+00	I	4.0E-01	C			1
Hydrogen Sulfide	6/4/7783					3.0E-03	I	2.0E-03	I			1
Lead and Compounds	7439-92-1											1
Tetraethyl Lead	78-00-2					1.0E-07	I					1

			Screening Levels									
RAGS Part E ABS	Csat	Residential Soil		Industrial Soil		Residential Air		Industrial Air		Tapwater		MCL
	mg/kg	mg/kg	key	mg/kg	key	$\mu g/m^3$	key	$\mu g/m^3$	key	$\mu g/l$	key	$\mu g/l$
	1.1E+05	6.1E+04	nc	6.1E+05	nc	3.2E+04	nc	1.4E+05	nc	2.2E+04	nc	
		1.4E+08	nc	6.0E+08	nc	1.0E+02	nc	4.4E+02	nc			
		3.1E+01	nc	4.1E+02	nc					1.5E+01	nc	6.0E+00
		3.9E+01	nc	5.1E+02	nc					1.8E+01	nc	
		7.0E+01	nc	9.2E+02	nc					3.3E+01	nc	
		3.1E+01	nc	4.1E+02	nc					1.5E+01	nc	
		3.1E+01	nc	4.1E+02	nc	2.1E-01	nc	8.8E-01	nc	1.5E+01	nc	
		1.5E+04	nc	1.9E+05	nc	5.2E-01	nc	2.2E+00	nc	7.3E+03	nc	2.0E+03
	2.0E+03	1.1E+00	ca*	5.6E+00	ca*	3.1E-01	ca	1.6E+00	ca*	4.1E-01	ca	5.0E+00
	2.6E+02	3.9E+03	sat	5.1E+04	nc					1.8E+03	nc	
	6.9E+02	7.7E-02	ca*	3.9E-01	ca*	8.1E-02	ca*	4.1E-01	ca*	1.6E-01	ca*	
	4.8E+02	2.5E-01	ca	1.3E+00	ca	1.6E-01	ca	8.2E-01	ca	2.0E-01	ca	5.0E+00
		7.5E+03	nc	9.1E+04	nc	1.5E-01	nc	6.4E-01	nc	3.7E+03	nc	
		1.2E+05	nc	1.5E+06	nc					5.5E+04	nc	
						2.0E-04	ca*	1.0E-03	ca*	1.1E+02	nc	
		2.3E+02	nc	1.4E+03	ca**	2.0E-04	ca	1.0E-03	ca			
		3.1E+03	nc	4.1E+04	nc					1.5E+03	nc	1.3E+03
0.10		3.1E+03	nc	3.1E+04	nc					1.8E+03	nc	
0.10		3.1E+03	nc	3.1E+04	nc					1.8E+03	nc	
0.10		3.1E+02	nc	3.1E+03	nc					1.8E+02	nc	
	3.1E+02	2.2E+03	sat	1.1E+04	nc	4.2E+02	nc	1.8E+03	nc	6.8E+02	nc	
	1.2E+02	7.2E+03	sat	3.0E+04	nc	6.3E+03	nc	2.6E+04	nc	1.3E+04	nc	
	1.4E+03	3.4E-02	ca	1.7E-01	ca	4.1E-03	ca	2.0E-02	ca	6.5E-03	ca	5.0E-02
	8.5E+02	1.9E+02	nc	7.8E+02	nc	2.1E+02	nc	8.8E+02	nc	3.9E+02	nc	
	1.9E+03	4.5E-01	ca	2.2E+00	ca	9.4E-02	ca	4.7E-01	ca	1.5E-01	ca	5.0E+00
	5.5E+02	5.7E+00	ca	2.9E+01	ca	9.7E-01	ca	4.9E+00	ca	1.5E+00	ca	7.0E+02
0.1		1.2E+05	nc	1.2E+06	nc	4.2E+02	nc	1.8E+03	nc	7.3E+04	nc	
		2.3E+02	nc	3.1E+03	nc	2.1E+00	nc	8.8E+00	nc	1.1E+02	nc	
		4.0E+02	nc									1.5E+01
0.1		6.1E-03	nc	6.2E-02	nc					3.7E-03	nc	

Continued

Contaminant											Toxicity and Chemical-specific Information	
		SFO		IUR		RfDo		RfCi				RAGS Part E
Analyte	CAS No.	$(mg/kg\text{-}day)^{-1}$	key	$(\mu g/m^3)^{-1}$	key	$(mg/kg\text{-}day)$	key	(mg/m^3)	key	voc	muta-gen	GIABS
Manganese (Diet)	7439-96-5					1.4E-01	I	5.0E-05	I			1
Manganese (Water)	7439-96-5					2.4E-02	I	5.0E-05	I			0.04
Methanol	67-56-1					5.0E-01	I	4.0E+00	C			1
Methoxyethanol, 2-	109-86-4							2.0E-02	I			1
Methyl Ethyl Ketone (2-Butanone)	78-93-3					6.0E-01	I	5.0E+00	I	V		1
Methyl tert-Butyl Ether (MTBE)	1634-04-4	1.8E-03	C	2.6E-07	C			3.0E+00	I	V		1
Nickel Refinery Dust	NA			2.4E-04	I							0.04
Nickel Soluble Salts	7440-02-0					2.0E-02	I					0.04
Nickel Subsulfide	12035-72-2			4.8E-04	I							0.04
Anthracene	120-12-7					3.0E-01	I			V		1
Naphthalene	91-20-3					2.0E-02	I	3.0E-03	I	V		1
Phenol	108-95-2					3.0E-01	I	2.0E-01	C			1
Phosphoric Acid	7664-38-2							1.0E-02	I			1
Styrene	100-42-5					2.0E-01	I	1.0E+00	I	V		1
Tetrachloro-ethylene	127-18-4	5.4E-01	C	5.9E-06	C	1.0E-02	I	2.7E-01	A	V		1
Toluene	108-88-3					8.0E-02	I	5.0E+00	I	V		1
Trichloroethane, 1,1,1-	71-55-6					2.0E+00	I	5.0E+00	I	V		1
Trimethylbenzene, 1,2,4-	95-63-6							7.0E-03	P	V		1
Xylene, Mixture	1330-20-7					2.0E-01	I	1.0E-01	I	V		1
Xylene, p-	106-42-3							7.0E-01	C	V		1
Xylene, m-	108-38-3					2.0E+00	H	7.0E-01	C	V		1
Xylene, o-	95-47-6					2.0E+00	H	7.0E-01	C	V		1
Zinc (metallic)	7440-66-6					3.0E-01	I					1
Zinc Phosphide	1314-84-7					3.0E-04	I					1

Key: I = IRIS; P = PPRTV; A = ATSDR; C = Cal EPA; H = HEAST; W = WHO; S = see User Guide Section 5;
L = see User Guide on lead; M = mutagen; V = volatile; ca = cancer; ca* = where: nc SL < 100× ca SL;
ca** = where nc SL < 10× ca SL; nc = noncancer; max = concentration may exceed Ceiling Limit
(see User Guide); sat = concentration may exceed Csat (see User Guide); SSL values are based on DAF = 1

RAGS Part E ABS	Csat (mg/kg)	Residential Soil (mg/kg)	key	Industrial Soil (mg/kg)	key	Residential Air (µg/m³)	key	Industrial Air (µg/m³)	key	Tapwater (µg/l)	key	MCL (µg/l)
		1.8E+03	nc	2.3E+04	nc	5.2E-02	nc	2.2E-01	nc	8.8E+02	nc	
0.1		3.1E+04	nc	3.1E+05	nc	4.2E+03	nc	1.8E+04	nc	1.8E+04	nc	
0.1		2.8E+07	nc	1.2E+08	nc	2.1E+01	nc	8.8E+01	nc			
	2.8E+04	2.8E+04	sat	1.9E+05	nc	5.2E+03	nc	2.2E+04	nc	7.1E+03	nc	
	6.9E+03	3.9E+01	ca	1.9E+02	ca	9.4E+00	ca	4.7E+01	ca	1.2E+01	ca	
		1.4E+04	ca	6.9E+04	ca	1.0E-02	ca	5.1E-02	ca			
		1.6E+03	nc	2.0E+04	nc					7.3E+02	nc	
		6.9E+03	ca	3.5E+04	ca	5.1E-03	ca	2.6E-02	ca			
0.13		1.7E+04	nc	1.7E+05	nc					1.1E+04	nc	
0.13		1.5E+02	nc	6.7E+02	nc	3.1E+00	nc	1.3E+01	nc	6.2E+00	nc	
0.1		1.8E+04	nc	1.8E+05	nc	2.1E+02	nc	8.8E+02	nc	1.1E+04	nc	
		1.4E+07	nc	6.0E+07	nc	1.0E+01	nc	4.4E+01	nc			
	1.0E+03	6.5E+03	sat	3.8E+04	nc	1.0E+03	nc	4.4E+03	nc	1.6E+03	nc	1.0E+02
	1.8E+02	5.7E-01	ca	2.7E+00	ca	4.1E-01	ca	2.1E+00	ca	1.1E-01	ca	5.0E+00
	9.3E+02	5.0E+03	sat	4.6E+04	nc	5.2E+03	nc	2.2E+04	nc	2.3E+03	nc	1.0E+03
	6.8E+02	9.0E+03	sat	3.9E+04	nc	5.2E+03	nc	2.2E+04	nc	9.1E+03	nc	2.0E+02
	2.5E+02	6.7E+01	nc	2.8E+02	nc	7.3E+00	nc	3.1E+01	nc	1.5E+01	nc	
	3.0E+02	6.0E+02	sat	2.6E+03	nc	1.0E+02	nc	4.4E+02	nc	2.0E+02	nc	1.0E+04
	4.5E+02	4.7E+03	sat	2.0E+04	nc	7.3E+02	nc	3.1E+03	nc	1.5E+03	nc	
	4.4E+02	4.5E+03	sat	1.9E+04	nc	7.3E+02	nc	3.1E+03	nc	1.4E+03	nc	
	3.0E+02	5.3E+03	sat	2.3E+04	nc	7.3E+02	nc	3.1E+03	nc	1.4E+03	nc	
		2.3E+04	nc	3.1E+05	nc					1.1E+04	nc	
		2.3E+01	nc	3.1E+02	nc					1.1E+01	nc	

Exposure Parameters

Target cancer risk	1E-06	TR
Target Hazard Quotient	1.0	THQ
Body weight, adult (kg)	70	BW_adult
Body wt, age 1–6 (kg)	15	BW-child
Default skin surface area for soil contact, adult resident (cm^2/day)	5700	SA_adult
Default skin surface area for soil contact , child (cm^2/day)	2800	SA_child
Default skin surface area for soil contact, adult worker (cm^2/day)	3300	SA_work
Default adherence factor, adult resident (mg/cm^2)	0.07	AF_adult
Default adherence factor, child (mg/cm^2)	0.20	AF_child
Default adherence factor, adult worker (mg/cm^2)	0.20	AF_work
Dermal absorption in soil (non-volatile organics)	0.10	ABS_org
Averaging time (years of life)	70	AT
Air breathed (m^3/d)	20	IRA_adult
	10	IRA_child
Drinking water ingestion (l/d)	2	IRW_adult
	1	IRW_child
Volatilization factor – water (l/m^3)	0.5	VF_W
Volatilization factor – soil (m^3/kg)		VF_S
Particulate emission factor (m^3/kg)	1.3E+09	PEF
Soil ingestion – adult resident (mg/d)	100	IRS_adult
Soil ingestion – child age 1–6 (mg/d)	200	IRS_child
Soil ingestion – adult worker (mg/d)	100	IRS_work
Exposure frequency (d/yr)	350	EF_R
Exposure duration, age 1–6 (yr)	6	ED_C
Exposure duration, adult (yr)	30	ED_A
Exposure duration, lifetime (yr)	70	ED_L
Residential age-adjusted factors for carcinogens only		
Ingestion factor for soils ([mg/yr]/[kg/d]) See text	114	IFS_adj
Skin contact factor for soils ([mg/yr]/[kg/d]) See text	361	SFS_adj
Inhalation factor ([m^3/yr]/[kg/d]) See text	11	InhF_adj
Ingestion factor for water ([l/yr]/[kg/d]) See text	1.1	IFW_adj
Exposure frequency, adult worker (d/yr)	250	EF_work
Exposure duration, adult worker (yr)	25	ED_work

Index